International Technology Transfer to Developing Countries

Kamal Saggi

COMMONWEALTH SECRETARIAT

This report is part of the Commonwealth Economic Paper Series
published by the Economic Affairs Division of the Commonwealth Secretariat
Commonwealth Secretariat
Marlborough House, Pall Mall
London SW1Y 5HX, United Kingdom

© Commonwealth Secretariat, 2004

Views and opinions expressed in this paper are the responsibility of the authors and
should in no way be attributed to the institutions to which they are affiliated.

All rights reserved. No part of this publication may be reproduced, stored in a retrieval
system, or transmitted in any form or by any means, electronic or mechanical, including
photocopying, recording or otherwise without the permission of the publisher.

Designed and published by the Commonwealth Secretariat
Printed in Britain by Formara Ltd.

Wherever possible, the Commonwealth Secretariat uses paper sourced from
sustainable forests or from sources that minimise a destructive impact on the environment.

ISBN 0-85092-795-1   Price: £8.99

Web site: http//www.thecommonwealth.org

Contents

Part I: International Technology Transfer: National Policies, International Negotiations and Multilateral Disciplines — 1

1 Introduction — 3

2 Trade, Knowledge Spillovers and Growth — 7
2.1 Knowledge Spillovers via Trade — 7
2.2 Empirical Evidence on Knowledge Spillovers — 9
2.3 Trade in Capital Goods — 10

3 The Role of Foreign Direct Investment — 13
3.1 Multinational Firms and Technology Transfer — 14
   A. Demonstration Effects — 15
   B. Labour Turnover — 16
   C. Vertical Technology Transfer — 17
3.2 Empirical Evidence on Horizontal Spillovers from Foreign Direct Investment — 20
3.3 Foreign Direct Investment and Economic Growth — 24

4 National Policies — 27
4.1 Trade Policy — 27
4.2 Policy on Foreign Direct Investment — 28
   A. Restrictions on Foreign Direct Investment — 28
   B. Investment Incentives — 33
4.3. Intellectual Property Rights Protection — 39

5 Multilateral Rules and Disciplines — 45
5.1 The TRIMS Agreement — 45
5.2 The TRIPS Agreement — 47

6 Policy Lessons — 49

Part II: Encouraging Technology Transfer to Developing Countries: The Role of the WTO — 55

1 Introduction — 57
1.1 Asymmetric Information — 58
1.2 Market Power — 58
1.3 Positive Externalities — 59

2 Trade as a Channel of Technology Transfer — 61
2.1 Research on Growth Effects of Trade — 61
2.2 Capital Goods Trade — 62

3 The Role of Foreign Direct Investment — 65
3.1 Multinational Firms and Technology Transfer — 65
3.2 Spillovers from Foreign Direct Investment — 66

4 National Policies — 69
4.1 Policies Towards Foreign Direct Investment — 69

5 Multilateral Rules and Disciplines — 75
5.1 TRIPS and ITT — 75
5.2 ITT and the Mandate of the WTO — 78
5.3 The Role of the WTO in Resolving Information Problems — 80
5.4 Curtailing Market Power — 81
5.5 The TRIMS Agreement — 82

References — 84

Tables
1.1: Global Exports of Capital Goods as a Percentage of Total Exports (1975–1996) — 52
1.2a: FDI Inward Stock, 1980–2000 ($ billion) — 52
1.2b: FDI Inward Stock as a Percentage of GDP — 52
1.3: Receipts of Royalties and Licenc e Fees, 1985–97 ($ million) — 53

Figure
1.1: FDI Inflows as a Percentage of GDI (Low- and Middle-income Countries) — 53

I

International Technology Transfer:
National Policies, International Negotiations
and Multilateral Disciplines

1
Introduction

Most countries lag behind the technology frontier in at least some industries and therefore confront the issue of how best to bridge the technology gap vis-à-vis the rest of the world. For developing countries, this issue is doubly critical: not only do they lag further behind relative to other countries, but they also face more stringent resource constraints that make the use of inefficient production technologies especially problematic. While most governments have always wanted to catch up technologically with the rest of the world, there has been wide variation in countries' actual economic performance. How does one explain this variation? In recent years, why has it only been some countries (such as South Korea and Japan) that have succeeded in catching up technologically with the rest of the world? Answers to such questions require a thorough understanding of the process of international technology transfer (ITT). As might be expected, ITT is a complex phenomenon – it occurs via a variety of channels and is influenced by a multitude of factors. The purpose of this report is to provide a comprehensive overview of the *economics* of ITT. Since many governments have tended to take a pro-active role toward the various channels of ITT, the report discusses both the role of market forces, and national and multilateral policies.

While the report focuses on international technology transfer, it does not contend that countries can (or should) rely solely on foreign research and development (R&D). In fact, all countries need to conduct some indigenous R&D in order to improve their technological capabilities. Given that technology can be imported, why must a country invest in domestic R&D? Following classical trade theory, should developing countries not simply purchase technology from other countries that have comparative advantage in R&D? There are at least two responses to this viewpoint. First, the prescription of specialisation based on comparative advantage applies only under a stringent set of assumptions, many of which are not supported by empirical evidence. For example, new technologies are rarely produced under conditions of perfect competition and the market for technology is plagued by transactions costs that stem (partly) from the presence of asymmetric information between buyers and sellers. Involvement in R&D by potential buyers of new technologies can facilitate exchange of technology by lowering transactions costs. As will be argued below, these aspects of the market for technology need to be taken into account to properly evaluate the historical policy initiatives of countries such as South Korea and Japan. A second reason for conducting domestic R&D is that technological change is a dynamic phenomenon and technology acquisition is not a one-time decision but an ongoing process.

Thus, a dichotomous 'make-or-buy' choice does not adequately capture the complexity of the process of ITT. The decision not to invest in local R&D may not only increase the transactions costs of ITT but may also force a country to bear such costs for a long time.

Technological change, while an interesting phenomenon in its own right, is important primarily because of its potential impact on productivity and economic growth. As a result, the subject has always invited keen interest from academicians and policy-makers alike. Of particular interest within this general theme has been the relationship between ITT and openness to international trade and foreign direct investment (FDI). While the notion of openness is easy to grasp intuitively, its empirical implementation has not always been straightforward.[1] Despite the ongoing debate in the literature regarding the effect of increased openness on economic growth, economists are generally supportive of greater openness.[2] In fact, a common view among them is that traditional analyses may understate the true cost of protectionism: being static in nature, most such analyses do not capture the dynamic costs of trade protection. Underlying this view is the notion that trade, FDI and interaction among countries in various other forms help improve the global allocation of resources and encourage transmission of technology across national borders.

A survey of relevant research conducted in this report indicates that this optimistic view is backed by some empirical evidence. But the evidence is not as clear-cut as proponents of greater openness would like. In fact, even at the theoretical level, it has not been easy to provide models that capture the link between openness and productivity in a way that yields empirically testable results. Future research can benefit from increased communication between theory and data since there appears to be a disjunction between the theoretical and empirical literature. Broadly speaking, empirical research has not shed enough light on the *mechanisms* via which openness helps improve productivity, while theoretical research that deals with the underpinnings of this connection has eluded empirical implementation.

Theoretical models of trade and FDI frequently lead to ambiguous welfare conclusions because introducing dynamics in an interesting fashion often requires multiple departures from the neoclassical model of perfect markets. Externalities and imperfect competition play a central role in the new dynamic models, and such distortions can easily lead to perverse results. However, it is also not the case that 'anything can happen' if a closed economy opens up to free trade. In fact, the theoretical literature suggests that the *scope of knowledge spillovers* is a crucial determinant of the effect of trade on productivity and growth (Grossman and Helpman, 1995). The empirical evidence on this issue has been mixed: some studies discover that knowledge spillovers have a limited geographical scope, whereas others find the opposite. The ambiguous nature of this empirical evidence begs the following question: what factors determine the scope of knowledge spillovers? The scope of knowledge spillovers must be partly determined by the interaction between innovators (potential suppliers of technology) and those agents that seek access to newer technologies through imitation, technology licensing and other forms of collaboration with innovators. In other words, a fair amount of technology transfer must indeed be endogenous. Since technology transfer is likely to be an important channel via which trade and FDI contribute to productivity, this report discusses the complex inter-relationships between trade, FDI and technology transfer in some detail.

To set the stage, it is useful to go back to Paul Romer's illuminating discussion of the special properties of knowledge as an economic good. His seminal paper on technological change (Romer, 1990) is built on the premise that knowledge is a *non-rival* good: the use of a certain type of knowledge by one person does not preclude another from using it. However, this does not mean that technology, and knowledge in general, can be acquired at zero cost. If technology transfer entailed no costs, the room for fruitful policy intervention with respect to assimilation of foreign technology would be quite limited because any technology transfer that would yield even a minutely positive return to any agent would take place automatically.[3] The non-rival nature of knowledge only implies that if two agents are willing to pay the cost of adopting a new technology, they can do so without interfering with each other's decisions. Much empirical evidence indicates that ITT carries significant resource costs (Teece, 1977; Mansfield and Romeo, 1980; Ramachandran, 1993).

Suppose the resource costs needed for acquisition of foreign technology have been incurred by a developing country. An important issue is: how does a technology newly introduced into an economy (say by a multinational firm) subsequently diffuse throughout the rest of the economy? The presence of trade barriers across countries, as well as international differences in market conditions and policy environments, implies that technology diffusion within a country should be considerably easier than diffusion across national borders. For example, the mobility of labour is severely constrained only at the international level (exceptions include contact with consultants and the return of foreign-educated nationals). Thus, labour turnover across firms may be crucial for driving technology diffusion within a country, and may not play any role in ITT. In this report, significant attention is given to the role of technology licensing, imitation and FDI in the process of ITT and its subsequent diffusion in the host country.

A major aim of this report is to help identify the role played by policy in facilitating ITT. The range of relevant policies is clearly quite large. To limit the scope of the discussion, only policies related to trade, FDI and intellectual property rights are discussed. Given the central questions of interest, the literature on FDI and intellectual property rights is discussed in greater detail than trade policy. A brief discussion of the major policy implications of the report is contained in a separate section at the end.[4]

2
Trade, Knowledge Spillovers and Growth

Traditional economic theory teaches that expansion in world trade yields efficiency gains by improving the global allocation of resources. The more interesting question is whether trade also yields dynamic efficiency gains by improving productivity growth in the world. To answer this question, it is natural to examine how trade affects technological change and global diffusion of new technologies.

2.1 Knowledge Spillovers via Trade

Standard neoclassical growth models assume costless ITT by positing a common production function across countries. The fact that chosen production techniques vary across countries is not evidence against the neoclassical view: such dispersion in production techniques will naturally result from differences in factor prices (that may be caused by the lack of similarity in factor endowments across countries). As Parente and Prescott (1994) note, the issue is whether firms in different countries can access the global pool of technologies at the same cost. They emphasise that barriers to technology adoption can be a key determinant of international differences in per capita income. According to this view, some countries make it inherently costlier for their firms to adopt modern technologies, thereby retarding economic development. Following this logic, increased openness may lead to more growth by lowering the barriers to technology adoption.

In contrast to neoclassical models that stress capital accumulation, the new growth theory assigns a central role to technological change and the accumulation of human capital (Lucas, 1988). Romer (1990), Grossman and Helpman (1991), Aghion and Howitt (1990) and Segerstrom, Anant, and Dinopoulos (1990) are among the pioneers of R&D based models of economic growth. These models formalise the Schumpeterian notion of 'creative destruction' and are built around the idea that entrepreneurs conduct R&D to profit from monopoly power that results from innovation.[5] Grossman and Helpman (1991) provide a unifying framework for two widely used strands of R&D based endogenous growth models: the varieties model that builds on foundations laid by Dixit and Stiglitz (1977), Ethier (1982) and Romer (1990), and the quality ladders model developed by Aghion and Howitt (1990), Segerstrom, Anant, and Dinopoulos (1990) and Grossman and Helpman (1991).

A discussion of the main assumptions underlying R&D based growth models is useful in shedding light on the relationship between trade and ITT. In a closed economy, growth is sustained in the varieties model through the assumption that the creation of new products expands the knowledge stock, which then lowers the cost of innovation. As more products are invented, both the costs of inventing new products and the profits of subsequent innovators are lower because of increased competition (no products disappear from the market in this model). In contrast, the quality ladders model assumes that consumers are willing to pay a premium for higher quality products. As a result, firms always have an incentive to improve the quality of products. The important *assumption* that sustains growth in this model is that every successful innovation allows all firms to study the attributes of the newly invented product and then improve on it. Patent rights restrict a firm from producing a product invented by some other firm, but not from using the knowledge (created by R&D) that is embodied in that product. Thus, as soon as a product is created, by assumption the knowledge needed for its production becomes available to all; such knowledge spillovers ensure that anyone can try to invent a higher quality version of the same product.[6]

R&D based growth models contain an important insight: since new products result from new ideas, trade in goods can help transmit embodied knowledge internationally. Of course, trade in ideas can take place *without* trade in goods. Rivera-Batiz and Romer (1991) analyze two different models (the lab equipment model and the knowledge-driven model) of endogenous growth in order to highlight the role of trade in goods versus trade in ideas. The general conclusion of their analysis is that trade in either goods or ideas can increase the global rate of growth if such trade allows a greater exploitation of increasing returns to scale (in the production of goods or ideas) by expanding market size.

Multi-country models of endogenous growth have two strands: those that study trade between identical countries and those that have a North–South structure. Although knowledge spillovers are central to both, technology transfer is a central feature only of North–South models. Prominent early works include Krugman (1979), Rivera-Batiz and Romer (1991) and Grossman and Helpman (1991). North–South models that emphasise the product-cycle nature of trade have been particularly useful for understanding ITT and merit some further discussion. These product-cycle models assume that new products are invented in the North and that, due to lower relative wages in the South, Southern firms can successfully undercut Northern producers by succeeding in imitating Northern products. A typical good is initially produced in the North until either further innovation (in the quality ladders model) or successful Southern imitation (in both the varieties model and the quality ladders model) makes profitable production in the North no longer feasible. Consequently, either production ceases (due to innovation) or shifts to the South (due to imitation). Thus, prior to imitation, all products are exported by the North, whereas post-imitation they are imported, thereby completing the cycle. These models capture technology-driven trade and have been generalised to consider technology transfer more explicitly.

Neither FDI nor licensing (choices available to innovators for producing in the South) were considered in the early variants of these models.

What do R&D based models of growth imply about the effect of trade on productivity and growth? An important conclusion of this line of research is that much hinges on whether knowledge spillovers are *national* or *international* in nature (Grossman and Helpman, 1995). If knowledge spillovers are international, these models endorse the view that trade is an engine of growth. However, when knowledge spillovers are national in scope, perverse possibilities can arise. Note that this perspective is more relevant for North–North models of trade because international knowledge spillovers (of one form or another) are *assumed* in North–South product-cycle models of trade, where the South is modelled as a pure imitator.[7]

What factors can help account for the explosive growth of countries like Hong Kong, South Korea, and Taiwan? Some economists argue that the accumulation of resources has driven economic growth in these countries (Young, 1995). Others argue that improvement in productivity (driven partly through trade) has played a large role (Nelson and Pack, 1999). However, even if capital accumulation were the driving force, why did it take place at such a high rate? What kept the returns to capital accumulation so high? Perhaps technology transfer (again, partly through trade) kept the marginal product of capital from falling and kept investment rates high (Nelson and Pack, 1999; Hsieh, 2002).

2.2 Empirical Evidence on Knowledge Spillovers

What does the empirical evidence say about the scope of knowledge spillovers? At a casual level, the frequent agglomeration of R&D intensive industries (such as in Silicon Valley) appears to suggest that spillovers are primarily local. Branstetter (2001) uses data on high-technology US and Japanese manufacturing firms from 1977–89 to check for the presence of knowledge spillovers by estimating the impact R&D of other firms had on the patents earned by a firm. He finds that knowledge spillovers are primarily national rather than international. However, other studies, especially those using macro data, do not reach the same conclusion.[8] Several such studies find that R&D activity in a country is not strongly correlated with productivity growth, suggesting that the benefits of R&D in one country spill over substantially to other countries. For example, Eaton and Kortum (1996) find that more than 50 per cent of the growth in some OECD countries derives from innovation in the United States, Germany and Japan. Yet Eaton and Kortum also report that distance inhibits the flow of ideas between countries, whereas trade enhances it. Coe and Helpman (1995) and Coe, Helpman and Hoffmaister (1997) argue that international R&D spillovers are substantial and that trade is an important channel for such spillovers. Using estimates of international R&D spillovers from these two studies, Bayoumi *et al.* (1999) simulate the impact of changes in R&D and in exposure to trade on productivity, capital, output and consumption in a multi-country model. Their simulations indicate that R&D can affect output not only directly, but also indirectly by stimulating capital investment. Incidentally, this finding is also of interest for the debate regarding the Asian growth 'miracle'.

Keller (1996) casts some doubt on the results of Coe and Helpman (1995) by generating results similar to theirs for randomly generated trade weights. However, a paper by Coe and Hoffmaister (1999) argues that Keller's 'random' weights are actually not random. When truly random weights are used, estimated international R&D spillovers are non-existent, implying that they only arise in the data when the actual trade weights of countries are used. Lichtenberg and Potterie (1998) also provide supporting evidence in favour of the hypothesis that the more open to trade a country is, the more it benefits from foreign R&D. Schiff *et al.* (2002) argue that when a country (A) imports a good from another country (B); there is no reason why A should only benefit from the R&D stock of B because B's knowledge stock is greater than its own R&D stock due to its trade with other countries. In other words, bilateral trade between two countries should result in indirect knowledge spillovers that arise due to the trade of each country with rest of the world. Using industry level data, Schiff *et al.* show that such indirect knowledge spillovers are actually larger than the standard spillovers estimated in the literature. Incidentally, their analysis also shows why Keller's so-called random weights work as well as they do: once indirect spillovers are allowed for, the difference among OECD countries in total foreign R&D stock is much smaller that that implied by the direct R&D measures used by Coe and Helpman. As a result, the distribution of trade weights ceases to play much of a role. The provocative implication of Schiff *et al.* is that a country can enjoy large knowledge spillovers by trading with only a few others as long as its trading partners are trading with a large number of countries.

2.3 Trade in Capital Goods

In principle, trade in both consumption and capital goods can contribute to technology transfer and the empirical studies discussed above typically utilise a country's imports of all goods while attempting to measure knowledge spillovers through trade. For example, when a country imports a manufactured consumption good (such as an automobile), local firms can absorb some technological know-how by simply studying the design and the engine of the imported automobile. While such attempts at reverse engineering are no doubt important, they probably contribute less to technology transfer than trade in capital goods (such as machinery and equipment) that are used in the production of other consumption goods. Xu and Wang (1999) plausibly argue that trade in capital goods is more relevant than total trade for measuring knowledge spillovers because capital goods have higher technological content than consumption goods.

Overall, capital goods trade is a prominent part of world trade and its importance has increased over time. In 1975, capital goods (defined as machinery and transport equipment) made up approximately 23 per cent of total world trade; in 1996 this proportion was over 30 per cent. During the period 1975–96, worldwide exports of capital goods as a percentage of GDP increased from about 4.2 per cent to approximately 7 per cent. In 1996, approximately 30 per cent of capital goods exports were destined for developing countries. Although the developing country share of imports of capital goods has increased over time, this increase has not been substantial (it was 28.9 per cent in 1980 compared with 30.8 per cent in 1996). Furthermore, developing countries are heavy importers of capital goods and some 85 per cent of the imports of machinery and transport equipment into developing countries come from developed countries (Mazumdar, 2001). Within the OECD countries, there is significant cross-country variation in the magnitude of imports of capital goods. In 1983–90, more than 50 per cent of US imports from other OECD countries comprised capital goods; for Japan this ratio was only 25 per cent (Xu and Wang, 1999). Such variation in the data suggests that using total trade to measure the degree of knowledge spillovers across countries might lead to erroneous conclusions.

Xu and Wang (1999) show that although the volume of capital goods trade helps to explain cross-country differences in total factor productivity, trade in all other goods does not. This result fits well with the finding by De Long and Summers (1991) that investment in machinery and equipment has a strong association with growth. Mazumdar (2001) takes the argument one step further. He notes that due to the presence of trade barriers, investment in imported equipment might generate more growth than investment in local equipment since the true opportunity cost of imported equipment is lower than that of domestically produced equipment which is produced under trade protection. His empirical analysis supports this argument and also shows that imported equipment matters more for developing countries than developed ones, perhaps because their economies are more open in general.

3
The Role of Foreign Direct Investment

Although the increase in world trade has received significant attention, the role FDI has played in the explosion of world trade is often under-appreciated. Today, intra-firm trade, that is trade between subsidiaries and headquarters of multinational firms, may account for one-third of total world trade. The importance of FDI can also be gauged from the fact that sales of subsidiaries of multinational firms now exceed worldwide exports of goods and services. In 1998, the total estimated value of foreign affiliate sales in the world was $11 trillion, whereas the value of global exports was $7 trillion (UNCTAD, 1999). Thus, FDI is the dominant channel through which firms serve customers in foreign markets.

Much of the flows of FDI occur between industrial countries (as does most intra-industry trade). For example, during 1987–92 industrial countries attracted $137 billion of FDI inflows a year on average; developing countries attracted only $35 billion, or slightly more than 20 per cent of global FDI inflows. Yet developing countries are becoming increasingly important host countries for FDI, especially because of the large-scale liberalisation undertaken by formerly communist countries such as China. During 1996 and 1997, over 40 per cent of global FDI flows went to developing countries (UNCTAD, 1999).[9] The recent surge in capital flows to developing countries, of which FDI has been a significant part, is also reflected in the fact that approximately 32 per cent of the total stock of FDI today is in developing countries (UNCTAD, 2001). It is worth noting that because of their smaller size, FDI is of relatively greater importance to developing countries. In 1999, the total inward stock of FDI as a percentage of GDP was almost 28 per cent in developing countries, compared with less than 15 per cent in industrial countries.[10]

For the purposes of this report, the role of FDI as a channel for transferring goods and services internationally is of secondary concern. Instead, the main issue of interest here is the role of FDI as a channel of technology transfer and its impact on productivity. Yet some discussion of the impact of FDI on market structure and competition is in order. Multinationals are creatures of market imperfection; they arise because markets are not always the most efficient means of exploiting intangible assets (such as technology) in international markets. Almost invariably, the very assets that lie behind the emergence of multinationals also create market power. Thus, while writing formal models of multinational firms, one has to abandon the widely prevalent model of perfect competition. A well accepted stylised fact is that the presence of multinationals is positively correlated with market concentration (Markusen, 1995). Of course, this is not say that restricting the entry of multinationals would make markets more competitive in developing countries. On the contrary, the restrictions faced by multinationals are likely to worsen market concentration in many international markets.

3.1 Multinational Firms and Technology Transfer

Multinational firms play a crucial role in ITT. For example, in 1995 over 80 per cent of global royalty payments for international transfers of technology were made from subsidiaries to their parent firms (UNCTAD, 1997). In general, technology payments and receipts have risen steadily since the mid-1980s, reflecting the importance of technology for international production. The data also indicate the importance of FDI for international trade in technology. During 1985–97 between two-thirds and nine-tenths of technology flows were intra-firm in nature. Furthermore, the intra-firm share of technology flows has increased over time. Of course, royalty payments only record the explicit sale of technology and do not capture the full magnitude of technology transfer through FDI relative to technology transfer via imitation, trade in goods and other channels.

Yet another confirmation of the strong role FDI plays in transmitting technology internationally comes from the inter-industry distribution of FDI. It is well known that multinational firms are concentrated in industries that exhibit a high ratio of R&D relative to sales and a large share of technical and professional workers (Markusen, 1995). In fact, it is commonly argued that multinationals rely heavily on intangible assets, such as superior technology, for successfully competing with local firms that are better acquainted with the host country environment.

By encouraging inward FDI, developing countries hope not only to import more efficient foreign technologies, but also to improve the productivity of local firms via technological spillovers to them.[11] Not surprisingly, a large literature tries to determine whether host countries enjoy spillovers from FDI. It is important to be clear about the meaning of the word *spillover*. A distinction can be made between pecuniary externalities (that result from the effects of FDI on market structure) and other pure externalities (such as the facilitation of technology adoption) that may accompany FDI. A strict definition of spillovers would count only the latter, and this is the definition employed here. In other words, if FDI spurs innovation in the domestic industry by increasing competition, this need not be viewed as a spillover from FDI because this effect works through the price mechanism.

The central difficulty in measuring spillovers is that spillovers do not leave a paper trail; they are externalities that the market fails to take into account. Nevertheless, several studies have attempted the difficult task of quantifying spillovers. But what are the potential channels through which such spillovers may arise? A more difficult question is whether it is even reasonable to even expect spillovers to occur from FDI.

Multinationals have much to gain from preventing the diffusion of their technologies to local firms (except when technologies diffuse vertically to potential suppliers of inputs or buyers of goods and services sold by multinationals – see below).

At the micro level, the literature suggests the following potential channels of spillovers:

- **Demonstration effects:** Local firms may adopt technologies introduced by multinational firms through imitation or reverse engineering;
- **Labour turnover:** Workers trained or previously employed by the multinational may transfer important information to local firms by switching employers, or may contribute to technology diffusion by starting their own firms;
- **Vertical linkages:** Multinationals may transfer technology to firms that are potential suppliers of intermediate goods or buyers of their own products.

A. Demonstration Effects

In its simplest form, the demonstration effect argument states that exposure to the superior technology of multinational firms may lead local firms to update their own production methods. The argument derives from the assumption that it may simply be too costly for local firms to acquire the necessary information for adopting new technologies if they are not first introduced in the local economy by multinationals (and hence demonstrated to succeed in the local environment). Incidentally, the demonstration effect argument relates well to the point made by Parente and Prescott (1994) that trade may lower the costs of technology adoption.

Clearly, *geographical proximity* is a vital part of the demonstration effect argument. As noted earlier, empirical evidence on the geographical scope of R&D spillovers is mixed. However, studies that reach optimistic conclusions with respect to the international nature of R&D spillovers typically involve data from industrial countries and therefore may not be applicable to developing countries. Geographical proximity may indeed be crucial for developing countries that are not as well integrated into the world economy and that have fewer alternative channels for absorbing technologies.

The main insight of the demonstration effect argument is that FDI may expand the set of technologies available to local firms. If so, this is a potential positive externality. However, a mere expansion in choices need not imply faster technology adoption, especially if domestic incentives for adoption are also altered due to the impact of FDI on market structure. FDI may expand choices, but it generally also increases competition. The industrial organisation literature on market structure and innovation does not provide an unambiguous answer to this question. A rough conclusion is that a monopolist has a stronger incentive to invest in R&D that yields innovations that complement existing technology, whereas competitive firms have stronger incentives to invest in innovations that replace existing technology.

Suppose that FDI does lower the cost of technology adoption and leads to faster adoption of new technologies by local firms. Does that imply that, relative to trade (i.e. in a scenario where foreign firms export to the domestic or world market), inward FDI necessarily improves productivity in the local economy? A point to keep in mind is that technology diffusion may strengthen the competitors of the foreign firms. Foreseeing the consequences of such diffusion, foreign firms may alter the very terms of their original technology transfer. For example, a foreign firm may choose to transfer technologies of lower quality when there is a risk of leakage to local firms.[12]

In Wang and Blomström's (1992) duopoly model with differentiated goods, a multinational transfers technology to its subsidiary given that the local firm can learn from the new technologies introduced. Learning occurs both through costless technology spillovers, as in the contagion effects that Findlay (1978) first emphasised, as well as through the local firm's costly investments. An interesting implication of Wang and Blomström's model is that technology transfer through FDI is positively related to the level of the local firm's investment in learning. This result suggests that multinationals respond to local competition by introducing newer technologies faster.

B. Labour Turnover

Although researchers have extensively studied direct imitation and reverse engineering as channels of inter-firm technology diffusion, they have tended to neglect the role of labour turnover. Labour turnover differs from the other channels because knowledge embodied in the labour force moves across firms only through the physical movement of workers. The relative importance of labour turnover is difficult to establish because it would require tracking individuals who have worked for multinationals regarding their future job choices and then determining their impact on the productivity of new employers. Few empirical studies attempt to measure the magnitude of labour turnover from multinationals to local firms. There appear to be *no* empirical studies that attempt to measure the role such turnover plays in improving productivity in local firms.

The available evidence on labour turnover itself is mixed. For example, although Gershenberg's (1987) study of Kenyan industries finds limited evidence of labour turnover from multinationals to local Kenyan firms, several other studies document substantial labour turnover from multinationals to local firms. UNCTAD (1992) discusses the case of the garment industry in Bangladesh (see also Rhee, 1990). Korea's Daewoo supplied Desh (the first Bangladeshi firm to manufacture and export garments) with technology and credit. Thus, Desh was not a multinational firm in the strict sense; rather, it was a domestic firm that benefited substantially from its connection with Daewoo. Eventually, 115 of the 130 initial workers left Desh to set up their own firms or to join other newly established garment companies. The remarkable speed with which the former Desh workers transmitted their know-how to other factories clearly demonstrates the role labour turnover can play in technology diffusion.

Pack (1997) discusses evidence documenting the role of labour turnover in disseminating the technologies of multinationals to local firms. For example, in the mid-1980s, almost 50 per cent of all engineers and approximately 63 per cent of all skilled workers who left multinationals left to join local Taiwanese firms. By contrast, Gershenberg's study of Kenyan industry reports smaller figures: of the 91 job shifts studied, only 16 per cent involved turnover from multinationals to local firms.

In order to synthesize these empirical findings the cross-country variation in labour turnover rates itself requires an explanation. One possible generalisation is that in countries such as South Korea and Taiwan, local competitors are less disadvantaged relative to their counterparts in many African economies, thereby making labour turnover possible. Thus, the ability of local firms to absorb technologies introduced by multinationals may be a key determinant of whether labour turnover occurs as a means of technology diffusion in equilibrium (see Glass and Saggi (2002c) for a formal model). Furthermore, the local investment climate may be such that workers who wish to leave multinationals in search of new opportunities (or other local entrepreneurs) find it unprofitable to start their own companies, implying that the only alternative opportunity is to join existing local firms. The presence of weak local competitors probably goes hand in hand with the lack of entrepreneurial efforts because both may result from the underlying structure of the economic environment.

Labour turnover rates may vary at the industry level as well. Casual observation suggests that industries with a fast pace of technological change (such as the computer industry in Silicon Valley) are characterised by very high turnover rates relative to more mature industries. Therefore, cross-country variation in labour turnover from multinationals could simply stem from the global composition of FDI; developing countries are unlikely to host FDI in sectors subject to rapid technological change.

C. *Vertical Technology Transfer*

For some time, analysts have recognised that multinationals may benefit the host economy through the backward and forward linkages that they generate. However, merely documenting extensive linkages between multinationals and local suppliers or buyers is an insufficient basis for arguing that net benefits accrue to the local economy as a result of FDI. Rodriguez Clare (1996) develops a formal model of linkages and shows that multinationals improve welfare only if they generate linkages over and beyond those generated by the local firms they displace. The relevant question here is whether the generation of linkages is expected to result in productivity improvements and/or technology diffusion.

Vertical technology transfer has been documented to occur when firms from industrialised countries choose to buy the output of firms in many Asian economies in order to sell it under their own name (Hobday, 1995). For example, companies such as Radio Shack and Texas Instruments have commissioned firms in developing countries to produce components or entire products which are then sold under the retailer's name. Rhee, Ross-Larson and Pursell (1984), summarising the results of extensive interviews in Korea in the late 1970s report that:

> *The relations between Korean firms and the foreign buyers went far beyond the negotiation and fulfillment of contracts. Almost half of the firms said they had directly benefited from the technical information foreign buyers provided: through visits to their plants by engineers or other technical staff of the foreign buyers, through visits by their engineering staff to the foreign buyers, through the provision of blueprints and specifications, through information on production techniques and on the technical specifications of competing products, and through feedback on the design, quality and technical performance of their products* (p. 61).

The knowledge transfers involved were multi-faceted: not only manufacturing knowledge was transferred but exact sizes, colours, labels, packing materials and instructions to users. It has also been found that in the later 1970s, many importing firms from industrialised countries maintained very large staffs in countries such as Korea and Taiwan who spent considerable time with their local manufacturers assisting them in meeting their specifications (Keesing, 1982). Motivated by this evidence, Pack and Saggi (2001) developed a model that explores the interdependence between production of manufactures in developing countries and their marketing into industrialised country markets. In their model, a buyer from an industrialised country can transfer technology to producers in a developing country in order to outsource production. Since firms in developing countries often lack the ability to successfully market their products internationally, technology leakage in the developing-country market actually *benefits* the industrial-country firm since it increases competition among the developing-country suppliers. An interesting implication of their analysis is that fully integrated multinational firms may be more averse to technology diffusion than firms that are involved in international arms-length arrangements.[13]

More recent evidence regarding vertical technology transfer is provided by Mexico's experience with the *maquiladora* sector and its automobile industry. Mexico started the *maquiladora* sector as part of its Border Industrialisation Programme designed to attract foreign manufacturing facilities along the US-Mexico border. Most *maquiladoras* began as subsidiaries of US firms that shifted labour-intensive assembly operations to Mexico because of its low wages relative to the US. However, the industry evolved over time and the *maquiladoras* now employ sophisticated production techniques, many of which have been imported from the US. In the automobile industry, one of Mexico's most dynamic sectors, FDI resulted in extensive backward linkages: within five years of initial investments by US firms, there were hundreds of domestic producers of parts and accessories. US firms and other multinational firms transferred technology to these Mexican suppliers: industry best practices, zero defect procedures, production audits+, etc. were introduced to domestic suppliers, thereby improving their productivity. As a result of increased competition and efficiency, Mexican exports in the automobile industry boomed. Thus, although direct competitors of multinational firms may not realise technological benefits (as evidenced by Aitken, Harrison and Lipsey 1996), suppliers of intermediate goods are likely to benefit substantially.

In a broader study, Batra and Tan (2002) use data from Malaysia's manufacturing sector to study effect of multinationals on inter-firm linkages and productivity growth during 1985–95. Their results show that not only are foreign firms more involved in inter-firm linkages than domestic firms, but also that such linkages are associated with technology transfer to local suppliers. Such technology transfers were found to have occurred through worker training and the transmission of knowledge that helped local suppliers improve the quality and timeliness of supply. Smarzynska (2002) examines backward linkages and technology spillovers using data from the Lithuanian manufacturing sector during the period 1996–2000. She finds that firm productivity is positively affected by a sector's intensity of contacts with multinational customers but not by the presence of multinationals in the same industry. Thus, her results support vertical spillovers from FDI but not horizontal ones. Furthermore, she finds that vertical spillovers occur only when the technological gap between domestic and foreign firms is moderate.

Blalock (2001) uses a panel dataset from Indonesian manufacturing establishments to check for the same effects. He finds strong evidence of a positive impact of FDI on productivity growth of local suppliers, showing that technology transfer from multinationals did indeed take place. He also plausibly suggests that since multinationals tend to source inputs that require relatively simple technologies relative to the final products they themselves produce, local firms that produce such intermediates may be in a better position to learn from multinationals than those that compete with them.

Even more interesting is the possibility that such vertical transfers, when accompanied by spillovers, may lead to interaction between upstream and downstream multinational firms that encourages industrial development. Markusen and Venables (1999) develop a model that abstracts from technology spillovers but focuses on the pecuniary externalities that accompany vertical linkages and result in industrial development. Pack and Saggi (2001) argue that buyers in industrial country markets benefit from technology diffusion among potential suppliers in developing countries because such diffusion increases competition and lowers the prices of intermediate goods. Their analysis implies that fully integrated multinational firms may be more averse to technology diffusion than firms that are involved in international arms-length arrangements.

3.2 Empirical Evidence on Horizontal Spillovers from Foreign Direct Investment

Early efforts to find horizontal spillovers from FDI proceeded by relating the inter-industry variation in productivity to the extent of FDI (Caves, 1974; Globerman, 1979; Blomström and Persson, 1983; Blomström, 1986). By and large, these studies find that sectors with a higher level of foreign involvement (measured by the proportion of the labour force in the industry employed by foreign firms or the extent of foreign ownership) tend to have higher productivity, higher productivity growth or both of these. The fact that these studies involve data from different countries (Australia in the Caves study, Canada in Globerman and Mexico in Blomström) lends a strong degree of robustness to this positive correlation between the level of foreign involvement and local productivity at the sector level.

Of course, correlation is not causation and, as noted by Aitken and Harrison (1999), this literature may overstate the positive impact of FDI on local productivity. Investment may have been attracted to the more productive sectors of the economy instead of being the cause of the high productivity in such sectors. In other words, the studies ignore an important self-selection problem. Both trade and FDI help to ensure an efficient allocation of global resources by encouraging investment in those sectors in which an economy enjoys comparative advantage. In this sense, the argument made by Aitken and Harrison is almost necessarily implied by traditional trade theory. However, if trade protection encourages investment in sectors in which a host economy does not enjoy comparative advantage, trade protection may be welfare reducing. This possibility was relevant for countries that sought to industrialise by following a strategy of import substitution.

Nevertheless, only plant-level studies can control for the self-selection problem that may plague industry-level studies. Taking the argument a step further, the self-selection problem may also arise in plant-level studies: the more productive plants may be the ones that attract foreign investment. For example, Clerides *et al.* (1998) find support for the self-selection hypothesis in the context of exporting. They find that the more productive firms self select into exporting. However, if plant-level studies fail to find a significant relationship between foreign involvement and productivity, the self-selection problem might not be important. It might be important if foreign firms seek out plants with low productivity and bring them up to par with the more efficient local plants. In this case, there might be no significant productivity differential between foreign and local firms. This argument seems far-fetched, but it could make sense as follows. Suppose local plants with very low productivity are relatively undervalued by local agents because the skills (technology and modern management) needed to make them competitive are in short supply locally. In this scenario, such plants would be attractive to foreign investors who can, through their technology, generate productivity improvements that simply cannot be achieved by local agents.

What do empirical plant-level studies find with respect to spillovers from FDI? Haddad and Harrison's (1993) study was the first to employ a comprehensive dataset at the level of the individual firm over several years. The data came from an annual survey of all manufacturing firms in Morocco. An important finding of this study was that foreign firms exhibited higher levels of total factor productivity (TFP), but their rate of TFP growth was lower than that for domestic firms. As the authors note, at first glance such a finding suggests that perhaps there was some sort of convergence between domestic and foreign firms. However, this was not the case. Although there was a level effect of foreign investment on the TFP of domestic firms, such an effect was missing for the growth rate of the TFP of domestic firms. In addition, when sectors were divided into high and low tech, the effect of FDI at the sector level was found to be more positive in low-tech sectors. The authors interpret this result as indicative of the lack of absorptive capacity on the part of local firms in the high-tech sector, where they may be further behind multinationals and unable to absorb foreign technology.

Aitken et al. (1996) employ a somewhat different approach to measuring spillovers from FDI. The idea behind this study is that technology spillovers should increase the marginal product of labour and this increased productivity should show up as higher wages. The study employs data from manufacturing firms in Venezuela, Mexico and the United States. For both Mexico and Venezuela, a higher share of foreign employment is associated with higher overall wages for both skilled and unskilled workers. Furthermore, royalty payments to foreign firms from local firms are highly correlated with wages. Most importantly, the study finds no positive impact of FDI on the wages of workers employed by domestic firms. In fact, the authors report a small negative effect for domestic firms, whereas the overall effect for the entire industry is positive. These findings differ from those for the United States, where a larger share of foreign firms in employment is associated with both a higher average wage as well as higher wages in domestic establishments. Putting the findings of Aitken et al. (1996) into the context of previous work, it is clear that wage spillovers (from foreign to domestic firms) are associated with higher productivity in domestic plants. Conversely, the absence of wage spillovers appears to accompany the existence of productivity differentials between domestic and foreign firms.

Using annual census data for more than 4000 Venezuelan firms, Aitken and Harrison (1999) provide another recent test of the spillover hypothesis. Since each plant was observed over a period of time, the self-selection problem of past sector-level studies (i.e. that FDI goes to the more productive sectors) could be avoided in their study. The authors find a positive relationship between foreign equity participation and plant performance, implying that foreign participation does indeed benefit plants that receive such participation. However, this own-plant effect is robust only for small plants, that is, those plants that employ fewer than 50 employees. For larger plants, foreign participation does not result in any significant improvement in productivity relative to domestic plants. More interestingly, Aitken and Harrison find that productivity in domestic plants *declines* with an increase in foreign investment. In other words, the authors find evidence of *negative spillovers* from FDI and suggest that these could result from a market stealing effect. That is, foreign competition may have forced domestic firms to lower output and thereby forgo economies of scale. Note that if loss in output is large enough, local plants may have lower productivity despite enjoying some sort of technology spillovers. Nevertheless, on balance, Aitken and Harrison find that the effect of FDI on the productivity of the entire industry is weakly positive. They also note that similar results are obtained for Indonesia, except that the positive effect on own plants is stronger, whereas the negative effect on domestic plants is weaker, suggesting a stronger overall positive effect.

Djankov and Hoekman (2000) also find a negative spillover effect of FDI on purely domestic firms in Czech industry. Interestingly, however, when joint ventures are excluded from the sample and attention is restricted to the impact of majority-owned foreign affiliates (that is, FDI) on all other firms in an industry (including joint ventures), the negative effect loses statistical significance. The authors report that survey questionnaires reveal that joint venture firms invest significantly more in new technologies than do purely domestic firms. They suggest that purely domestic firms might lack the ability to absorb the technologies introduced by foreign firms (due to their lower R&D efforts).

In a recent paper, Haskel *et al.* (2002) use plant-level panel data for all UK manufacturing from 1973 through 1992 to re-examine the issue of spillovers from FDI. As the authors note, there can be little doubt that local firms in the UK possess sufficient absorptive capacity to benefit from the introduction of newer technologies by multinationals. So if spillovers do not materialise, this cannot be attributed to the limitations of domestic firms. Across a wide range of specifications, the authors find that there are positive spillovers from FDI at the industry level.[14] More precisely, they find that a 10 per cent increase in foreign presence in a UK industry raises the total factor productivity of that industry's domestic plants by about 0.5 per cent.[15]

To recapitulate, several studies have cast doubt on the view that FDI generates positive spillovers for local firms. But such findings need not imply that host countries have nothing significant to gain (or that they are bound to lose) from FDI. Domestic firms should be expected to suffer from an increase in competition; in fact, part of the benefit of inward FDI is that it can help weed out relatively inefficient domestic firms. Resources released in this process will be put to better use by foreign firms with superior technologies, efficient new entrants (both domestic and foreign) or some other sectors of the economy. However, such reallocation of resources cannot take place instantaneously. Existing studies of spillovers may not cover a long enough period to be able to accurately determine how FDI affects turnover rates (entry and exit). Furthermore, their design limits such horizontal studies because they cannot clarify linkages and spillovers that may result from FDI in industries other than the one in which FDI occurs.

A challenge facing the optimistic view regarding technology spillovers from FDI is to explain how such spillovers can ever be in the interest of the multinational firms. Clearly, under most circumstances, multinationals would prefer to limit diffusion in the local economy. In fact, the heart of the theory that seeks to explain the emergence of multinationals is that such firms are able to compete successfully with local firms precisely because they possess superior technologies, management and marketing. Why, then, would multinationals not take action to ensure that such advantages do not diffuse to local competitors?

Part of the answer must lie in the fact that such actions are costly and may even entail externalities between multinationals. Suppose a costly action (such as litigation in local courts to enforce protection of intellectual property rights) can indeed help limit the loss of knowledge capital. A difficulty arises if all potential multinationals benefit, whereas the costs fall only on the one which takes legal action. Thus, the public good nature of such actions suggests that developing countries hosting multinationals may expect the rivalry among such firms to result in some degree of technology diffusion. Of course, the preceding argument also somewhat overstates the case: some loss of knowledge will result despite all precautions. Nevertheless, it is beyond dispute that multinationals can take action to limit diffusion, and while they are making their decisions regarding where and how to set up subsidiaries, the expected costs of technology diffusion will enter into their calculation of profit maximisation.

It is worth emphasising that the entry of multinationals may benefit host countries even if it fails to result in spillovers for local firms. First, the preceding discussion suggests that spillovers to local firms that directly compete with the multinationals would be the most elusive of the benefits that host countries may expect to enjoy from FDI. Second, local agents other than domestic competitors of multinationals (for example local workers and local suppliers) may enjoy positive externalities from FDI. If so, the total effect of FDI on local welfare may be positive despite the lack of technology spillovers. Third, spillovers may be of an entirely different nature: local firms may enjoy positive externalities from foreign firms that make it easier for them to export. Such externalities may come about because better infrastructure (transportation, storage facilities and ports) emerges in regions with a high concentration of foreign exporters. Aitken, Hanson and Harrison (1997) provide direct evidence on this issue. They conducted a detailed study of 2104 manufacturing plants in Mexico. In their sample, 28 per cent of the firms had foreign ownership and 46 per cent of the foreign plants exported. Their major finding is that the probability of a Mexican-owned plant exporting is positively correlated with its proximity to foreign-owned exporting plants. Such spillovers may result from informational externalities and are more likely to lower fixed costs, rather than marginal costs, of production.

3.3 Foreign Direct Investment and Economic Growth

Regardless of the channel through which technology spillovers occur, the fact that FDI often involves capital inflows along with technology transfer implies that one would expect a positive impact of FDI on growth in the host country. Yet there are several important caveats to this assertion. First, a positive correlation between the extent of FDI and economic growth in cross-country regressions may simply reflect the fact that countries that grow faster attract more FDI. Thus, the causation could run from growth to FDI. Second, multinationals often raise the required capital in the host country, and in such a scenario capital inflows associated with FDI may not be substantial. An optimistic view of FDI would then look to technology transfer and/or spillovers as the mechanism through which FDI affects growth. Indeed, Romer (1993) argues that FDI can have a positive effect on growth in developing countries by helping them bridge the 'idea gap' with respect to industrial countries.

Glass and Saggi (2002a) examine the question of spillovers from FDI in a product-cycle growth model. In their North–South model, the demonstration/proximity argument is formalised as follows. Southern firms are assumed to be able to imitate multinationals located in the South at a lower cost than firms located in the North. However, multinational firms are also stronger competitors than firms that produce only in the North because multinationals produce in the same low-wage location as potential imitators. The model delivers the surprising result that a faster flow of FDI increases the aggregate rate of technology transfer to the South only if local firms lack the ability to imitate firms located in the North (that is, if geographical proximity is a prerequisite for imitation). If firms in the North can be imitated, FDI does not alter aggregate technology transfer because imitation focusing on firms located in the North slows down with a hastening of imitation targeting multinationals. Thus, their model takes a nuanced view of local absorptive capacity; it is useful to think of local capacity in terms of the different channels of technology transfer that may be accessible to a country.

Almost all theories of FDI and licensing have been developed in either static or partial equilibrium models. A few dynamic general equilibrium models explore the effect of FDI on growth, but these models have ignored the possibility of licensing. Glass and Saggi (2002b) develop a model of FDI that captures the internalisation decision and its implications for both the rate and magnitude of innovation. They also examine how policy interventions (taxes or subsidies to FDI) that alter the incentive to internalise production within the firm affect economic growth. They find that the ability of firms to switch modes from licensing to FDI in response to policy changes is vital for ensuring that a subsidy to FDI leads to faster economic growth.

In a comprehensive paper, Borensztein, De Gregorio and Lee (1998) utilise data on FDI flows from industrial countries to 69 developing countries to test the effect of FDI on growth in a cross-country regression framework. Their findings are as follows. First,

FDI contributes more to domestic growth than domestic investment, suggesting that it is indeed a vehicle of technology transfer. Second, FDI is more productive than domestic investment only when the host country has a minimum threshold stock of human capital. The latter finding is especially interesting because it clarifies when exactly FDI should be expected to effect growth.

In another empirical study, Balasubramanyam et al. (1996) use cross-section data from 46 developing countries to investigate the effect of inward FDI on growth in such countries. They report two main findings. First, the growth-enhancing effects of FDI are stronger in countries that pursue a policy of export promotion rather than import substitution, suggesting that the trade policy regime is an important determinant of the effects of FDI. Second, they find that in countries with export-promoting trade regimes, FDI has a stronger effect on growth than domestic investment. Both findings relate well to the results of Borensztein et al. (1998). The second finding may be viewed as a confirmation of the hypothesis that FDI results in technology transfer.

The findings of Borensztein et al. (1998) relate well to Keller (1996), who argues that mere access to foreign technologies may not increase the growth rates of developing countries. In his model, if a country's absorptive capacity (stock of human capital) remains unchanged, a switch to an outward orientation does not lead to a higher growth rate. Using a model quite different from Keller's, Glass and Saggi (1998) focus on the issue of the quality of technology transferred through FDI. They argue that investment in imitation by host-country firms generates the necessary knowledge (or skill) foundation for FDI, and thus factors that promote imitation can promote a higher-quality mix of FDI. Keller's model stresses that a country's stock of human capital effectively constrains its ability to take advantage of foreign technologies; Glass and Saggi (1998) emphasise that indigenous technological capability in an industry effectively constrains a country's ability to host foreign technology. Thus, they take a more micro-level view of the constraints on technology transfer than does Keller (1996), although both papers make similar points. For example, a country may have a fair amount of human capital in the aggregate, but may lack the technological sophistication to be able to host high-quality FDI in any particular industry.

Xu (2000) provides yet another confirmation of the argument that, in the absence of adequate human capital, technology transfer from FDI may fail to increase productivity growth in the host country. Using data on outward FDI from the United States to 40 countries, Xu measures the technology transfer intensity of multinational firms' affiliates by their spending on royalties and license fees as a share of their gross output. He finds that technology transfer from FDI contributes to productivity growth in more developed countries but not in less developed countries because the latter lack adequate human capital. Incidentally, as Xu notes, FDI may contribute to productivity growth due to reasons other than technology transfer. Thus, a statistically significant coefficient on some measure of FDI in a productivity growth equation does not necessarily imply that technology transfer is the mechanism through which FDI contributes to productivity growth. Xu estimates that, of the total effect of trade (through R&D spillovers) and FDI (through technology transfer) on productivity growth in industrial countries, 41 per cent is due to technology transfer. Xu's results for industrial countries confirm the findings of Barrell and Pain (1997), who find that FDI has a positive impact on technological change in West Germany and the UK.

Xu and Wang (2000) find that although capital goods trade serves as a channel of technology transfer among industrial countries, bilateral flows of FDI do not. However, Xu and Wang raise questions regarding these results because of the poor quality of the FDI data. More encouraging results regarding the role of FDI in technology transfer have been found by Lichtenberg and Potterie (2001).[16] Using data from the US, Japan and 11 European countries, they investigate the effect of FDI on total factor productivity of these countries. An attractive feature of their empirical approach is that they differentiate between inward and outward FDI and check whether FDI contributes to total factor productivity growth holding constant the role of trade in the same process. Their main result is that outward FDI flows and imports are simultaneous channels of ITT. However, they find that inward FDI does *not* contribute to the technological development of host economies. This last result resonates with the questions raised in this paper regarding the optimistic view of spillovers from FDI since multinational firms will generally not benefit from a diffusion of their technology to local competitors. Their results lead to the provocative conclusion that FDI flows between developed countries are primarily means by which firms keep up with technological change (and perhaps demand shifts) in host countries, while also serving local customers. It is important to note that Potterie and Lichtenberg (2001) do not study FDI flows from developed to developing countries so their results cannot shed light on whether inward FDI into developing countries (from developed ones) serves as a channel of technology transfer.

4
National Policies

In this section we discuss the effects of three types of policies on technology transfer:

- Trade policy
- FDI policy
- Policies toward intellectual property rights.

4.1 Trade Policy

Although the literature on trade policy is voluminous, it does not pay significant attention to the interaction between protection and technology transfer. In fact, most models treat the process of technology transfer in a rather rudimentary way, focusing instead on other aspects of the problem. Here, a few prominent examples of this line of research are discussed.

Miyagiwa and Ohno (1995) examine a domestic firm's incentives for technology adoption when a foreign rival has already adopted a superior technology. They assume that the cost of adoption decreases over time, and they examine how the nature (tariff versus quota) and the duration (temporary versus permanent) of trade protection influence the domestic firm's incentives for technology adoption. Their most interesting result is that temporary protection (that is, protection that is removed on successful adoption by the domestic firm) actually delays the date of technology adoption. In a related paper, Miyagiwa and Ohno (1999) show that if temporary protection is credible, it may indeed increase R&D relative to free trade. However, if the domestic firm expects that protection will be removed early should innovation occur before the pre-announced terminal date of protection, the firm will invest less in R&D under protection relative to free trade. Similarly, as first emphasised by Matsuyama (1990), if the domestic firm expects protection to be extended if there is no innovation by the terminal date, its investment incentives are marred by protection.

The literature also investigates the effect of trade protection in R&D based models of endogenous growth (see Grossman and Helpman, 1991 and 1995). As can be expected from models in which increasing returns, imperfect competition and externalities play a central role, the results depend on the details of a particular model and require careful interpretation. To the extent that one can draw a general conclusion from such a complex literature, it is that the literature does not provide an unconditional argument against trade protection. The conclusions hinge dramatically on the scope of knowledge spillovers: international knowledge spillovers strongly tilt the balance in favour of free trade, whereas national spillovers create a role for policy intervention that can combat path dependence resulting from a historical accident. For example, if productivity improvements depend solely on a country's own R&D, a case can be made for policies that ensure that industries in which such improvements occur at a rapid rate are not all located elsewhere.

Dinopoulos and Segerstrom (1999) develop a specific-factor variant of the quality ladders model of endogenous growth without scale effects. They examine the consequences of contingent protection, that is, tariffs imposed on imports whenever domestic firms lose their technological leadership to foreign firms who successfully innovate over them. Their approach is interesting because protection in the real world is usually not marginal (for example, anti-dumping duties may be levied on foreign firms with the explicit goal of providing sufficient relief to domestic industry). Interestingly, Dinopoulos and Segerstrom (1999) find that tariffs that allow domestic firms to capture the domestic market are positively related to the global rate of technological change in the short run.

Grossman and Helpman (1991) also analyze the effects of tariff protection in a two-country quality ladders model. Unlike Dinopoulos and Segerstrom, Grossman and Helpman analyze only tariffs that are too small to allow domestic firms to capture the market. Both models assume Bertrand competition on the product market, so that a low-quality firm can monopolise the market only if a tariff of sufficient magnitude is imposed on higher-quality imports. A small tariff can extract rents from foreign firms, but fails to protect domestic firms that have been innovated over by foreign firms. It should be noted that Dinopoulos and Segerstrom assume that both countries adopt symmetric policies.

4.2 Policy on Foreign Direct Investment

There is no simple way of describing the policy environment that faces multinationals in developing countries. A roughly accurate statement is that while FDI in services markets faces a multitude of restrictions, FDI into the manufacturing sector is confronted with both restrictions and incentives, sometimes in the same country.

A. Restrictions on Foreign Direct Investment

In countries that historically emphasised import-substituting industrialisation – such as most of Africa, Latin America and south-east Asia – FDI was either completely prohibited or multinational firms had to operate under severe restrictions. In fact, even in countries where technology acquisition was a major policy objective, multinationals were rarely permitted to operate fully-owned subsidiaries; Japan, Korea and Taiwan all imposed restrictions on FDI at various points in time. In other words, 'outward-oriented' economies were *not* particularly keen on allowing multinational firms into their markets. Japan's Ministry of International Trade and Investment (MITI) played an active role in the country's acquisition of foreign technology. MITI limited competition between potential Japanese buyers, did not allow inward FDI until 1970, never greatly liberalised FDI and even insisted at times that foreign firms share their technology with local firms as a precondition for doing business in Japan. Ozawa (1974) provides a rich description of the role played by imported technology and local R&D (aimed at facilitating absorption of foreign technology) in Japan's economic development. What is the rationale behind policies that discourage FDI? Pack and Saggi (1997) argue that by prohibiting FDI and placing other restrictions on the conduct of multinationals, government policies in many countries may have effectively weakened the bargaining position of foreign firms. They note that in Japan, MITI restricted many local firms from participating as potential buyers for precisely this reason.

Restrictions on FDI were often accompanied by a more lax attitude towards other modes of ITT. To fully understand the effect of such restrictions, it is important to evaluate the broad FDI policy environment in the 'Japan-Korea' model. In contrast to their restrictive policies toward FDI, both Japan and South Korea aggressively encouraged licensing of foreign technology (Layton, 1982). Consider South Korea as an example: annual inflows of licensed technology into South Korea increased steadily during the 1970s and 1980s, whereas FDI inflows, which were always relatively low, stagnated during 1978–83 (Sakong, 1993). This slowdown of FDI into South Korea was partially a result of the restrictive FDI policies instituted by the Korean government during this period (Hobday, 1995). The experience of South Korea and Japan is fairly representative in one important respect: there are few, if any, examples of countries that encouraged FDI but restricted technology licensing. Sometimes policy also favours joint ventures relative to fully-owned subsidiaries of multinationals. For example, the Chinese government has been particularly interventionist in technology transactions and has encouraged FDI in the form of joint ventures. Although fully-owned subsidiaries of foreign firms are not prohibited from doing business in China, the policy environment favours joint ventures over such enterprises. Of course, an immediate reason for this might be that all such policies simply reflect protectionism. Large public firms or hitherto protected private firms who are unable to compete with multinationals may be able to secure protection through the political process. Hoekman and Saggi (2000) suggest that, although the motivation behind policies that discriminate between licensing, joint ventures and the establishment of wholly-owned subsidiaries is not easy to decipher, a plausible interpretation may be that such policies seek to maximise technology transfer to local agents while limiting the rent erosion that results from the entry of multinational firms.

To evaluate the argument that policy might be biased in favour of licensing to encourage local technological development, Saggi (1999) has developed a two-period model in which a foreign firm chooses between FDI and technology licensing. The key assumption of this model is that licensing results in greater transfer of know-how to the local firm than does FDI, under which the local firm must compete with the subsidiary of the multinational firm. The main result of this paper is that the local firm would have the strongest incentive for innovation if the foreign firm were to follow initial licensing by direct investment. However, in equilibrium, the foreign firm never adopts such a course of action. The implication of this analysis is that local incentives for R&D might be best encouraged by initially preferring licensing to FDI. However, the difficulty with this policy prescription is that it gives no clear idea about when to remove the restrictions on FDI. In fact, as is known from the literature on the credibility of trade protection, temporary restriction of FDI could very well become permanent: if it makes sense to restrict FDI today due to some underlying features of the local economy, it could very well make sense to restrict FDI in the future as well. Put simply, the problem confronting the policy-maker is: how to ensure that local incentives for innovation and other performance-enhancing activities are not hampered by the very policies that seek to encourage local industrial development?

The analytical objections to the policy preference for licensing and joint ventures notwithstanding, what does empirical evidence tell us about this issue? Using plant-level data for 1991 for all Indonesian establishments with more than 20 employees, Blomström and Sjoholm (1999) examine two important questions. First, do establishments with minority and majority ownership (that is, joint ventures versus wholly-owned subsidiaries) differ in terms of their (labour) productivity levels? Second, does the degree of technology spillovers vary with the extent of foreign ownership? The second question is especially relevant for our purposes. Blomström and Sjoholm (1999) obtain several interesting results. First, as in many other previous studies, they find that labour productivity is higher in establishments with foreign equity than in purely domestic firms. Second, the extent of total foreign production is positively associated with the productivity of domestic firms, suggesting some sort of spillover from FDI. And third, the degree of foreign ownership affects neither the productivity of firms that get foreign equity nor the extent of spillovers to the domestic sector. These results are interesting but one needs to keep in mind several things. First, the study only measures labour productivity and not total factor productivity of firms. Second, the study treats some important endogenous variables as exogenous. Third, its findings are rather puzzling: the degree of foreign participation does matter in that plants with no foreign investment are less productive. So how does it come about that those which receive more foreign investment are not more productive than those which receive less? Perhaps the results suggest some sort of threshold effects in which, beyond a certain degree of foreign ownership, additional foreign equity affects neither the productivity of those plants that receive foreign investment nor the degree of spillovers to local firms. Unfortunately, since the authors do not report the minimum level of foreign equity (for those plants that do receive foreign equity) in their sample, this argument cannot be evaluated empirically.

The focus on technology spillovers to local firms can also lead one to ignore a crucial issue: restrictions on FDI relative to licensing also need to account for the incentives of foreign firms regarding technology transfer under the two modes. Here, the evidence is less supportive of the Japan-Korea model. Several earlier studies document that technologies transferred to wholly-owned subsidiaries are of a newer vintage than licensed technologies or those transferred to joint ventures (Mansfield and Romeo, 1980; Kabiraj and Marjit, 1993; Saggi, 1996). Thus, by forcing multinationals to license their technologies, host countries might also be lowering the quality of technologies they receive.

Another policy issue is that many south-east Asian countries still do not allow free entry of multinational firms and often express preferences with regard to the type of FDI; that is, entry by Pepsi or Coke is viewed differently than entry by GM or Texas Instruments. Unfortunately, the literature provides little insight into such policies. Other than the standard argument that certain industries are able to secure greater protection than others, perhaps it may also be the case that spillovers to the local economy are higher under certain types of FDI. For example, it might be that domestic content protection policies involve more local firms and therefore generate greater spillovers. However, there is no formal model or empirical evidence that supports this position. In addition, this argument is closely related to the idea of industrial targeting in general, and the pitfalls of the government's ability to correctly identify 'high spillover' industries are well known.

Today, the most frequently observed policy restrictions on FDI are those on the number of foreign firms allowed to enter the market and on the extent of foreign ownership permitted. The pattern of these restrictions differs across countries and often across sectors within countries. For instance, consider policy in basic telecommunications services. At one end, in the Philippines, a high degree of competition co-exists with limitations on foreign equity partnership. Bangladesh and Hong Kong are examples of countries that have no limitations on foreign ownership, but both have monopolies in international telephony and oligopolies in other segments of the market. Pakistan and Sri Lanka have allowed limited foreign equity participation in monopolies to strategic investors, and deferred the introduction of competition for several years. Korea, however, is allowing increased foreign equity participation more gradually than competition. How might one explain the variation in these policies?

To shed light on this crucial question, Mattoo et al. (2003) model a foreign firm's choice between acquisition and direct entry when the degree of technology transfer is endogenously determined. Since this is the only formal model of technology transfer that addresses this important policy issue, we discuss it in some detail. Mattoo et al. show that when a foreign firm faces a high cost of technology transfer, it will generally prefer direct entry to acquisition – higher costs of technology transfer are associated with a smaller cost advantage over domestic firms and a higher acquisition price. A welfare-maximising domestic government, on the other hand, prefers acquisition because it leads to more technology transfer – the higher technology transfer under acquisition generally more than offsets its anti-competitive effect. Thus, when the cost of technology transfer is high, restricting direct entry and/or encouraging acquisition can help achieve a higher level of welfare in the host country. On the other hand, if the cost of technology transfer is low, the government prefers direct entry to acquisition. Under this scenario, direct entry is not only associated with a more competitive domestic market but also brings more technology transfer due to the foreign firm's stronger 'strategic' incentive to transfer technology. But, the foreign firm would rather acquire an existing firm. Not only does acquisition yield greater market power but also the acquisition price is small when the cost of technology transfer is low. In this scenario a restriction on foreign acquisition of an existing domestic firm can help induce direct entry. It is clear that for intermediate levels of technology transfer, both the government and the foreign firm prefer acquisition, and therefore there is no need for policy intervention.

By endogenising technology transfer and the mode of entry in an oligopolistic setting, the model in Mattoo et al. offers some interesting insights. First, their results show that when a foreign firm has different modes of entering the domestic market, its preferences need not align perfectly with those of the domestic economy. This divergence stems from the different objectives of the foreign firm and the government. Of course, such a divergence is almost implied whenever there exist certain market failures such as imperfect competition and asymmetric information. But as noted above, markets in which technology transfer is an important consideration are usually not competitive: multinational firms are found mostly in oligopolistic industries and they use their knowledge-based assets to compete with local firms who are usually better positioned in the market. The analysis in Mattoo et al. (2003) contributes to a better understanding of the implications of mode choice for technology transfer and welfare. It shows that certain types of policy restrictions on FDI might arise from attempts by local governments to improve local welfare in an environment of imperfect competition and costly technology transfer. Of course, it also shows that other types of policy restrictions can be quite counter-productive.

The policy implications of the analysis of Mattoo et al. should be treated with caution: they develop their results in a simple model with some special assumptions. For example, their analysis does not extend to a monopolistic competitive setting, where rents are dissipated by domestic or foreign entry into the host country. In such a set-up, the case for policy intervention vanishes: in the long-run equilibrium, the foreign firm and the government would be indifferent between the two modes of entry. One implication of their results is that FDI restrictions should be more commonly observed in markets where there are high barriers to entry. As noted earlier, FDI restrictions are indeed more common in services (telecom, utilities, banking, etc.)

than in manufacturing, where barriers to entry are generally lower, and for which most countries offer incentives to FDI rather than try to restrict it.

In addition to facing entry barriers and restrictions on the degree of ownership, multinationals often have to contend with explicit performance requirements. Two prominent examples of such policies are domestic content requirements and export performance requirements. We postpone the discussion of such policies until we comment on the role of the agreement on TRIMS (trade-related investment measures).

Despite the prevalence of various types of restrictions on FDI, multinationals do not necessarily face an entirely hostile environment in developing countries. In fact, many countries try to lure large multinational firms with investment incentives. Interestingly enough, it is not unusual to find investment incentives being offered in conjunction with performance requirements and other restrictions on FDI, perhaps to partially offset the negative impact of the latter on the likelihood of investment by multinationals. The schizophrenic nature of the overall policy environment reflects the guarded optimism with which many developing countries view the entry of multinational firms into their territory. We next discuss the impact of investment incentives.

B. Investment Incentives

Perceptions about multinational firms and their effects on host countries have undergone something of a transformation in the last 50 years or so. Almost all countries are now eager to attract FDI (particularly in the manufacturing sector) and many have concluded bilateral investment treaties (BITs) with important source countries. By 1999, over 1,600 BITs had been negotiated, compared to some 400 at the beginning of 1990 (UNCTAD, 1997). This explosion in the number of BITS can be seen as an expression of a relatively sanguine view of FDI. In fact, we have even stronger evidence of the optimistic view of FDI: approximately 116 countries take a pro-active approach towards FDI today and offer incentives to foreign investors (Moran, 1998). Incentives designed to lure in FDI can take the form of up-front subsidies deigned to help multinationals defray some of their fixed costs (financial incentives), tax holidays (fiscal incentives) and other grants. The main goal of such policies is to alter either the magnitude or the location of inward FDI. The use of incentives raises some important questions: (a) are there any economic grounds for the widespread use of investment incentives?; (b) do such incentives have the desired impact (i.e. lead to more FDI in countries that use them); and (c) does the use of incentives clash with other policy goals?

Basic economic theory tells us that it is optimal to subsidise an activity if it generates positive externalities – i.e. the activity benefits agents other than those directly involved in the activity itself. As is well known, activities that generate positive externalities tend to be under-provided by the market. Thus, market forces will generate too little FDI if such investment is accompanied by positive externalities. Such externalities could exist either between investors or from investors to other agents in the host country. Note, however, that if the dominant externalities are those that exist between potential investors, then it is unclear whether or not host countries ought to be subsidising inward FDI. However, investment incentives may be justified if host countries enjoy externalities from inward FDI. So the key question becomes: why might we expect positive externalities from FDI to host countries? Here, we have a rather classic candidate: countries may offer incentives to FDI assuming that increased FDI will generate greater technological spillovers for local firms.

As noted earlier, there exists a large literature that tries to determine whether or not host countries enjoy spillovers from FDI (see Section 1.3). Recall that several studies find that local competitors of multinationals do not enjoy technology spillovers while others find that they do. However, studies that have examined the impact of multinationals on their suppliers indicate that local suppliers benefit strongly from FDI. But again, the issue is not whether local suppliers gain from FDI or not. Rather, the issue is whether local suppliers enjoy benefits over and beyond the level that multinationals perceive them to. For example, if a multinational firm helps a local supplier lower its costs by 10 per cent and in turn enjoys a 10 per cent reduction in the price of the input it purchases from the supplier, then the multinational fully realises the benefit accruing to the supplier and there would be no positive spillovers. Of course, if the local supplier can secure some of the rents that come about from the cost reduction, the local economy will enjoy a positive externality that can potentially justify the use of investment incentives. Alternatively, if the process of cost reduction spills over from one local supplier to another, there can again be grounds for the use of investment incentives. Even in such cases, investment incentives will rarely be the first best policy on purely economic grounds (whereas they might be when all the political costs of alternative instruments are taken into account).

Consider now the possibility of potential positive externalities among investors. Lin and Saggi (1999) emphasise that such externalities can arise if investors are confronted with markets about which they have very little information. The recent remarkable shift in the policy environment in many developing and formerly communist countries implies that firms from industrialised countries have gained access to hitherto closed markets and many cheap locations of production for the first time. While the profit incentive for investing in such newly liberalised markets is clear, it is also the case that firms from industrialised countries have little experience in operating in such new environments. Add to this lack of experience the fact that FDI (especially greenfield FDI) is a complex undertaking, and it immediately follows that firms seeking to invest in these markets can learn valuable lessons from the successes and failures of other firms which have made similar decisions before them. FDI is a complex undertaking: it involves hiring foreign labour, setting up a new plant, meeting foreign regulations and developing new marketing plans. Such decisions can be made properly only with a considerable amount of information. In this context, decisions made by rival firms can lower a firm's fixed costs of investment by helping it to avoid mistakes.

Infrastructure spending, as well as technology transfer undertaken by early multinational investors, may lower similar expenditures of succeeding investors. For example, China required early foreign investors to assist in the development of local technology and human capital by training workers, setting up research institutes, etc. Again, such investments by early investors are likely to benefit later investors. Thus, there are good reasons to expect that positive externalities exist amongst firms investing in previously closed markets. Using a survey of Japanese firms planning investments in Asia, Kinoshita and Mody (1997) highlight the role private information plays in determining investment decisions. They emphasise that information regarding a variety of operational conditions (such as the functioning of labour markets, literacy and productivity of the labour force, timely availability and quality of inputs, etc.) may not be publicly available. In such a scenario, information is either gathered through direct experience or through the experience of others. Thus, present investment is a function of one's own investment in the past and of the behaviour of rival investors. Their empirical analysis indeed finds strong effects of own past behaviour and current rival behaviour regarding investments.

The implication of Kinoshita and Mody's findings and of the general presence of externalities between rival investors (be they informational in nature or not) is that since the benefits of their own investment may not be fully captured by early investors, too little FDI may be forthcoming or that there may be herd effects in that investment by one is followed by investment by others. Lin and Saggi (1999) provide a duopoly model in which the first firm to switch to FDI from exporting confers a positive externality on the subsequent investor by lowering its fixed cost of FDI. The presence of such externalities can lead to a 'wait-and-see' attitude on the parts of investors, thereby leading to investment delay which may be partially resolved via incentives.

There is a long tradition in the management literature which argues the prevalence of such 'follow-my-leader' approaches among multinational firms (Caves, 1996). The prevalence of such behaviour amongst multinationals suggests that an alternative case for the use of FDI incentives can be made on the basis of the oligopolistic nature of the markets within which FDI occurs. For example, consider Mexico's recent experience with FDI in its automobile industry. Initial investments by US car manufactures into Mexico were followed by investments not only by Japanese and European car manufacturers, but also by firms who made automobile parts and components (i.e. their suppliers). As a result, competition in the automobile industry increased at multiple stages of production, thereby improving efficiency. Such a pattern of FDI behaviour (i.e. investment by one firm followed by investment by others) reflects the strategic considerations involved in FDI decisions. Since multinational firms compete in highly concentrated markets, they are highly responsive to each other's decisions. An implication of this interdependence between competing multinationals is that a host country may be able to unleash a sequence of investments by successfully inducing FDI from one or two major firms. More broadly, if the local economy lacks a well-developed network of potential suppliers, multinational firms might be hesitant to invest and local suppliers may not develop because of lack of demand. In the presence of such interdependence, the development of an economy can be subject to a co-ordination problem that can partially be resolved by initiating investments from key firms. Similarly, several models in the economic geography and development literature generate low-level equilibrium traps. The existence of such traps provides a rationale for incentives in order to get over a critical mass (agglomeration) threshold required to attract firms to a location. Of course, co-ordination problems in industrial development are too big an issue to be tackled only by the use of investment incentives, but such policies can certainly help.

There exists another important social welfare argument in support of investment incentives. Competition between governments for FDI via the use of incentives can help to ensure that FDI goes to those locations where it is most highly valued. Of course, the use of incentives would be required only if investors lack some information about host countries that governments possess. If all agents are symmetrically informed, the most valuable location for a given investment project would be common knowledge. But when host country governments possess information that investors lack, investment incentives can be justified. For example, Bond and Samuelson (1986) argue that temporary tax holidays can act as an efficient signalling device: high productivity countries can signal their productivity to uninformed potential investors via tax holidays. It is rational for the country to make temporary tax concessions to an uninformed foreign investor since the initial loss in revenue can be recouped in the future: the investor is willing to tolerate high subsequent tax rates only in a high-productivity country. Thus, in the presence of asymmetric information, competition for inward FDI among countries may help to improve the allocation of capital across jurisdictions by ensuring that FDI moves to those countries where it has the highest social return. The deeper point here is that incentives actually help improve efficiency by resolving the distortion that arises from asymmetric information.

Having described potential arguments in favour of investment incentives, it is now time to ask whether such policies are effective. Many studies have concluded that incentives are not effective once the role of fundamental determinants of FDI has been taken into account (Caves, 1996; Moran, 1998). An implication of this finding is that incentives may end up as transfers to multinationals without influencing their location decisions. From an efficiency standpoint, if fiscal incentives fail to alter the pattern of FDI, they are not distortionary. In principle, if incentives are ineffective, there is no efficiency rationale prohibiting their use – it is in each country's self-interest not to offer them to investors. In such situations they are pure transfers from developing countries to multinationals, and developing countries should unilaterally choose not to use them. Of course, this argument assumes full information on the part of governments. In practice, investors may indicate to governments that incentives are required or else they will invest elsewhere. Not knowing the true location preferences of investors, host country governments might feel tempted to offer incentives, knowing that investment might occur even if they did not offer them. In cases where the investment would have occurred anyway, investment incentives can have a major distributional impact. Recent figures suggest that this is not a trivial concern. For example, in 1996, Mercedes Benz received a subsidy of $300 million from Alabama for establishing an auto plant: this amounted to a subsidy of $200,000 per employee (Moran, 1998). Similarly, Germany paid a subsidy of $6.8 billion to Dow Chemical (this amounts to an astounding $3,400,000 per employee!). It seems difficult to believe that the lifetime wages and benefits that employees receive at these chosen plants exceed their compensation at alternative employers by an amount equal to the subsidy per employee. In other words, one needs to factor in the opportunity cost of employment elsewhere when calculating the net benefit of the subsidy, and by this metric the case for the overpayment argument seems strong.

Even government officials themselves are often not convinced of the inefficacy of incentives, as illustrated by statements by a number of representatives in the WTO Working Group on Trade and Investment (WTO, 1998). To some extent this may reflect differences in views regarding what is meant by an incentive. It is important to distinguish between fiscal and financial incentives (which are usually firm-specific) and more general policies that promote business activity. That the latter matter a lot in attracting investment is uncontested. In a recent empirical analysis of the effect of US state-level policies on the location of manufacturing investment, Holmes (1998) found that the share of manufacturing in employment in states with a pro-business regulatory environment increased by one-third compared to a bordering state without one. Policies that encourage the adoption and adaptation of know-how and other such general incentives that apply across the board are important determinants of the economic fundamentals of an economy (for example effective enforcement of contracts, the absence of red tape, adequate infrastructure, and training and education programmes).

Is there is no evidence in favour of the efficacy of incentives? This is not the case. Barry and Bradley (1997) describe Ireland's experience with FDI. Both favourable policies (reduced taxes and trade barriers, and investment grants) as well as strong fundamentals (such as infrastructure and an educated labour force) seem to have played a role in attracting FDI to Ireland. The strong performance of the Irish economy since the mid-1980s is attributable both to strong fundamentals and significant FDI inflows into Ireland. Several studies find that fiscal incentives do affect location decisions, especially for export-oriented FDI, although they seem to play a secondary role (see

Guisinger et al., 1985; Hines, 1996; Devereux and Griffiths, 1998). However, fiscal incentives have been found to be unimportant for FDI geared towards the domestic market. This type of FDI has been found to be more sensitive to the extent to which it will benefit from import protection. Thus, a more sophisticated view of the efficacy of incentives may be in order: while they may be useful for attracting certain types of FDI, incentives do not seem to work when applied at a general level.

Does the fact that multinationals operate in oligopolistic environments raise some additional concerns regarding the use of incentives? The distribution of rents between governments, host country firms and large multinationals has always been a classic controversy. In contrast to industrialised countries, where two-way flows of FDI are large, developing countries are large net importers of FDI and it is precisely in developing countries that multinationals have been most controversial. When coupled with the expectation of technology spillovers to local firms, the use of investment incentives to multinationals becomes perplexing. The technology spillover view is that multinationals might help in the development of local industry. On the other hand, investment incentives to multinationals can put their local competitors at a disadvantage. Can the two views be reconciled? Perhaps the use of incentives imposes a short-run cost on local firms who may gain from foreign investment in the long run (see discussion of spillovers above).

Furthermore, a selective use of investment incentives can also have strategic consequences among foreign firms. This is a real possibility since multinationals are pervasive in markets with a high level of concentration. For example, an exporting foreign firm from a third country (or a local host firm) may find itself at a disadvantage with respect to a foreign firm that experiences a decline in its cost due to an investment subsidy. Thus, incentives can alter the distribution of rents amongst multinationals.

It is worth noting that the use of incentives for FDI is by no means restricted to developing countries. In fact, the absolute magnitude of such incentives is larger in industrialised countries, where they result from competition between jurisdictions within states. Of particular concern are incentives that reflect efforts by industrialised countries to retain or attract FDI that would be more efficiently employed in developing countries. Labour unions and groups representing the interests of local communities may oppose plant closures and efforts by firms to transplant facilities. Similar motivations underlie the use of trade policy instruments such as anti-dumping. It is important, therefore, to distinguish between locational competition that may be efficiency-enhancing and the use of investment and trade policies (such as anti-dumping) that alter the incentives for outward FDI. The latter policies are inherently inefficient because they protect industries that are no longer competitive and induce a variety of ancillary distortions that are well documented in the literature (Finger, 1993).

To recap, there are three potential rationales for the use of investment incentives.

First, FDI might result in positive externalities. Second, there might be asymmetric information between governments and investors. And third, investors' decision-making might be subject to interdependence that gives arise to co-ordination problems. Of course, there are also sound arguments against incentives: if effective, they can distort the pattern of investment and also cause major redistribution in favour of those that receive them at the expense of their local and foreign competitors. The irony here is that if they are ineffective, incentives cannot be distortionary at the same time. It is worth emphasising that the use of incentives may pit developing countries against each other in a bidding war. Such competition for investment will generally be to the detriment of the parties involved and may even lead to excessive payment to investors. Thus, there are valid reasons to question the use of investment incentives. Even if one accepts the notion that there is a solid economic rationale for providing incentives to FDI, empirical evidence regarding the efficacy of financial incentives in attracting FDI is ambiguous. Furthermore, given the difficulty of quantifying the positive externalities associated with inward FDI, determining the optimal incentive scheme is obviously very difficult. In other words, it is difficult to say that existing levels of incentives are optimal by any definition since there are no good estimates of the level of externalities involved. A recent paper that asks whether existing subsidies to FDI are justified by technological spillovers finds that this is not the case: even though there is evidence of spillovers in this study, the level of incentives is found to be too high (Haskel et al., 2002).

A final comment is in order. The ambiguous nature of the empirical evidence on the efficacy of incentives might stem from the fact that they are often used in conjunction with some sort of investment measures. In some cases, the latter may more than offset the attractiveness of incentives to investors, thereby leading to ambiguous empirical results.

4.3 Intellectual Property Rights Protection

Common sense suggests that if any set of policies affects ITT, it will be the host country's intellectual property rights (IPR) regime. The theoretical literature has often investigated the effect of IPR enforcement on technology transfer and FDI in several endogenous growth models. Other approaches also exist. For example, in a strategic partial equilibrium model, Vishwasrao (1995) argues that the lack of adequate enforcement of technology transfer agreements may encourage FDI relative to licensing. In her screening model, depending on the type of licensee, licensing may or may not lead to imitation. The trade-off between FDI and licensing is that FDI avoids the risk of imitation at the expense of higher production costs.

Given the goal of this report, I omit models in which technology transfer does not play a central role. Several of the papers are linked through their use of the two models used intensively by Grossman and Helpman (1991). Before turning to these, I discuss

Taylor's (1994) work because it differs from the other papers in that it employs a model of endogenous technological change with Ricardian features.

In a two-country model, Taylor examines two scenarios: one in which IPR enforcement is symmetric across the two countries (it applies to innovators regardless of country of origin) and one in which it is asymmetric (it protects only domestic innovators). Although Taylor conducts the analysis under the assumption of costless technology transfer and equal productivity in R&D in the two countries, his results hold even when these assumptions are dropped, making it possible to apply them to a North–South setting. A subtle qualification must be made: symmetric versus asymmetric treatment implies that both countries adopt one policy as opposed to another. Taylor's model does not analyze incentives for unilateral adoption of a symmetric policy. His major result is that asymmetric protection of IPR distorts the pattern of world trade and lowers the global rate of growth.

In a North–South model, interpreting the exogenous rate of imitation as a proxy for the level of Southern IPR enforcement and assuming innovation to be exogenous, Helpman (1993) shows that a decline in the intensity of imitation promotes FDI to the South. Krugman (1979) also addressed this issue, although his model has a greater degree of exogeneity than Helpman's. The major contribution of Helpman's work lies in providing the first detailed welfare analysis of IPR enforcement in the South (as measured by an exogenous decline in the rate of imitation) in a dynamic general equilibrium growth model. He shows that a strengthening of IPR protection is not in the interest of the South, and that a weak enforcement of IPR protection in the South may even benefit the North, provided the rate of imitation is not too fast. Lai (1998) extends the Helpman model to allow for FDI and shows that innovation is promoted along with FDI if the South strengthens its IPR protection. The common weakness of both models is that stronger IPR enforcement is modelled as an exogenous decline in the rate of imitation. Nevertheless, Helpman's model is a tour de force in that it clearly specifies the alternative channels through which a strengthening of Southern IPR protection affects Northern and Southern welfare.

Yang and Maskus (2001) study the effects of Southern IPR enforcement on the rate of innovation in the North, as well as on the extent of technology licensing undertaken by Northern firms. A key assumption in their model is that increased IPR enforcement increases the licenser's share of rents and reduces the costs of enforcing licensing contracts, thereby making licensing more attractive. Thus, both innovation and licensing increase as IPR protection in the South becomes stronger.

Glass and Saggi (2002a) provide an analysis of Southern IPR protection in a comprehensive product-cycle model of trade and FDI. In their model, Southern imitation targets both multinationals producing in the South and purely Northern firms producing in the North. They treat stronger IPR protection as an increase in imitation cost, perhaps stemming from stricter uniqueness requirements in the South. In their model,

FDI actually decreases with a strengthening of Southern IPR protection because an increase in the cost of imitation crowds out FDI through tighter Southern resource scarcity. Although products like books, videos and CDs receive a lot of publicity in relation to conflicts over IPR protection, imitating most products is not so simple (Pack and Westphal 1986). Empirical evidence indicates that imitation is indeed a costly activity for a wide range of high-technology goods, such as chemicals, drugs, electronics and machinery. For example, Mansfield, Schwartz and Wagner (1981) find that the costs of imitation average 65 per cent of the costs of innovation (and over 20 per cent for most products).

Less efficient imitation absorbs more resources, although the rate of imitation declines with a strengthening of Southern IPR protection. In addition, the contraction in FDI increases resource scarcity in the North: increased production leaves fewer resources for innovation, so the rate of innovation falls. It is worth emphasising that if strengthening Southern IPR protection increases the cost of imitation, targeting both firms producing in the North as well as multinationals producing in the South, Northern incentives for FDI (at the firm level) are basically unaffected.

It should be clear from the discussion so far that the theoretical literature does not give an unambiguous prediction regarding the effects of stronger Southern IPR protection on the extent of FDI and the rate of growth. Does empirical evidence help to resolve the issue? By and large, the literature has not explored the interaction between optimal policies in the two regions (for an exception, see Lai and Qiu, 1999).

Consider the effect of Southern IPR enforcement on FDI. Surveys of US multinational firms frequently find that such firms are more willing to invest in countries with stronger IPR protection (Lee and Mansfield, 1996). Mansfield (1994) reports survey evidence claiming that IPRs are important for location decisions in FDI, though the importance varies by type of investment. He also asked US firms for their perceptions about the weakness of IPRs in various partner countries, with these perceptions being based on whether IPRs impeded or enhanced decisions to undertake licensing and joint ventures. On the basis of this information, he found that the weakness of IPRs was significantly and negatively related to investment decisions across countries, suggesting that countries which strengthen their patent regimes could well attract additional FDI inflows.

How does the researcher reconcile the ambiguous predictions of the theoretical models with this empirical finding? There are two ways out. First, increased IPR enforcement can be asymmetrical in that firms that invest in a country may expect to have a greater influence in local courts relative to those that simply export. Second, imitation of firms located in the North may not be an option for local firms in some developing countries, as is assumed by some theoretical models. In such a scenario, any increase in IPR enforcement by the South will benefit multinational firms, thereby encouraging them to engage in FDI.

As Ferrantino (1993) notes, all the preceding models suffer from a fundamental problem: either FDI or licensing is the only channel through which Northern firms are allowed to produce in the South. A more complete treatment of FDI requires that Northern firms be given the option of transacting in technology through the market. What are the consequences of strengthening IPR protection in the South if Northern firms can choose between licensing and FDI? Does FDI increase with IPR enforcement or does such a change in policy encourage licensing by lowering the risk of opportunism in market transactions? The latter scenario is equally likely and studies that ignore the possibility of licensing (or joint ventures for that matter) are likely to overstate the effect of IPR enforcement on inward FDI. In fact, a more subtle analysis may be needed. Increased IPR enforcement by the South may indeed make it a more attractive location for production (thereby increasing FDI relative to exports). However, the technologies transferred for that purpose might flow through licensing rather than FDI, so that the net effect on technology transfer through FDI is ambiguous. Of course, aggregate technology transfer to the South may increase, although general equilibrium effects may also require qualifications of this conclusion (Glass and Saggi, 2002a).

Using data for 1982 on US exports and sales of overseas affiliates of US firms, Ferrantino (1993) presents a detailed cross-country study that attempts to identify the determinants of both exports and sales of multinational affiliates of US firms, as suggested by the gravity model. His analysis reveals many insights, but perhaps the most interesting finding is that the US firms export more to their affiliates in countries that have weak IPR regimes. Ferrantino (1993) suggests that this result may reflect attempts by US firms to limit technology leakage to their rivals abroad by confining production within the United States. This interpretation fits well with a central theme of this survey: multinational firms will adjust their strategies to optimise against policies and market conditions they face in various host countries, casting doubt on the conclusions of empirical (or theoretical) work that treats FDI as given.

Empirical evidence indicates that the level of IPR protection in a country also affects the composition of FDI in two different ways (Lee and Mansfield, 1996; Smarzynska, 1999). First, in industries for which IPRs are crucial (pharmaceuticals, for example), firms may refrain from investing in countries with weak IPR protection. Second, regardless of the industry in question, multinationals are less likely to set up manufacturing and R&D facilities in countries with weak IPR regimes, and more likely to set up sales and marketing ventures because the latter run no risk of technology leakage.

These studies present useful findings, but are unable to address perhaps the most central question of all: does a country's IPR regime affect its economic growth? Although there are several theoretical analyses of this question, empirical studies are scarce. One such study is Gould and Gruben (1996), which uses cross-country data on patent protection, trade regime and economic fundamentals. The study finds that IPR protection, as measured by the degree of patent protection, is an important determinant of economic growth. More interestingly, it finds that the effect of IPR protection is stronger for relatively open economies than it is for relatively closed ones. In other words, a strengthening of IPR protection is more conducive to growth when it is accompanied by a liberal trade policy.

A possible interpretation of this finding is that by increasing foreign competition trade liberalisation not only curtails monopoly power granted by IPRs, but also ensures that such monopoly power is obtained only if the innovation is truly global. If firms in other countries can export freely to the domestic market and have better products or technologies, a domestic patent is useless in granting monopoly power. Furthermore, that trade liberalisation alone can improve productivity. Using data from Mexican manufacturing firms, Tybout and Westbrook (1995) find that trade liberalisation is associated with higher rates of productivity growth. The results of Gould and Gruben (1996) show that IPR enforcement matters over and above trade orientation, and that they have mutually reinforcing effects.

Finally, what does the empirical literature tell us about the effect of IPR protection on trade? Theory shows us that asymmetric IPR protection across countries can distort the pattern of world trade; empirical evidence supports this result. Using bilateral trade data for manufactured goods from 22 exporting countries to 71 importing countries, Maskus and Penubarti (1995) find that within the group of large developing countries, the strength of IPR protection in the importing country (as measured by patent rights) exerts a significantly positive effect on bilateral manufacturing imports in many product categories. In other words, in such countries, weak IPR protection is indeed a barrier to the manufacturing exports of most OECD countries. Maskus (2000) provides an up-to-date discussion of the empirical evidence on the effects of IPR protection on trade and FDI. Smith (1999) updates the study by Maskus and Penubarti (1995), using data on exports from US states to 96 countries. She makes the interesting point that since countries with strong IPR protection also have sophisticated technological capabilities that facilitate local imitation of foreign technologies, within industrial countries there is an ambiguous relationship between the strength of IPR protection and the volume of trade.

5
Multilateral Rules and Disciplines

While there is no multilateral agreement that deals explicitly with technology transfer, several multilateral agreements of the WTO have direct bearing on ITT.

In so far as the General Agreement on Tariffs and Trade (GATT), or for that matter the General Agreement on Trade in Services (GATS), is concerned, its major objective of encouraging trade liberalisation can have only a positive effect on technology transfer. As discussed above, trade in goods (especially capital goods) plays a central role in diffusing technology internationally. Since there is nothing really controversial regarding the effect of GATT rules on technology transfer, this discussion will examine agreements on TRIMS and trade-related intellectual property rights (TRIPS). The TRIMS agreement is important because of the strong connection between technology transfer and FDI, while the TRIPS agreement deals explicitly with intellectual property rights.

5.1 The TRIMS Agreement

In the Uruguay Round, an agreement on TRIMS was negotiated. This agreement prohibits measures that are inconsistent with national treatment and the GATT ban on the use of quantitative restrictions. It contains an illustrative list of prohibited measures, including local content, trade-balancing, foreign exchange balancing and domestic sales requirements. Furthermore, it requires that all non-conforming policies be notified within 90 days of entry into force of the agreement and that these be eliminated within two, five or seven years, for industrialised, developing and least developed countries, respectively. The TRIMS agreement did not go much beyond existing GATT rules – it simply re-iterated the GATT national treatment principle and the prohibition of quantitative restrictions in the context of certain investment policies that are deemed to be trade-related.[17] The GATT has been a constraint on countries using TRIMs, and can be expected to become a more serious source of discipline in the future as Uruguay Round transition periods for developing countries expire.

What do we know about the economics of TRIMs? First, if domestic distortions and externalities from FDI are absent, then governments should allow for unfettered market transactions with respect to FDI. For example, under perfect competition, domestic content protection lowers welfare by raising the price of domestic inputs: the resulting benefits to input suppliers are outweighed by the costs incurred by final goods producers (Grossman, 1981). A rationale for policies restricting FDI can arise if there are domestic policy distortions or market failures. Since multinational firms typically arise in oligopolistic industries, the presence of imperfect competition in the host economy is an obvious candidate.[18] Analyses of content protection and export performance requirements under conditions of imperfect competition illustrate that the welfare effects of such policies need not be always negative (Hollander, 1987; Richardson, 1991 and 1993; Rodrik, 1987). However, the standard normative prescription applies: more efficient instruments can be identified to address the specific distortion at hand. For example, in the case of anti-competitive practices resulting from market power or collusion, appropriate competition policies need to be used. Similarly, domestic policy distortions such as tariffs should be removed at source. If the distortion is due to some type of market failure, an appropriately designed regulatory intervention is required. Further, such intervention needs to be applied on a non-discriminatory basis to both foreign and domestic firms. This approach is implicit in the WTO, which not only aims at progressive liberalisation of trade, but also imposes national treatment and most favoured nation (MFN) constraints on policies. The adoption of such principles entails a prohibition on the use of most trade-related investment measures.

The literature on TRIMs also notes that there may indeed be circumstances where, from the viewpoint of an individual country, its optimal second-best policy toward inward FDI has a restrictive flavour. However, such policies typically have a beggar-thy-neighbour effect. If all countries pursue such policies, the outcome will be inefficient from a world welfare point of view.[19] Under such circumstances, co-operation that involves agreement not to restrict FDI can then be Pareto improving. Alternatively, the situation may be zero-sum, in which case there are no gains from co-operation.

Surveys often find that investment measures require firms to take actions that they would have taken anyway. For example, a policy that requires firms to export is inconsequential if firms find it advantageous to export even in the absence of such a requirement. For example, surveys by the US Department of Commerce for 1977 and 1982 indicated that only 6 per cent of all the overseas affiliates of US firms felt constrained by TRIMs such as local content or export performance requirements, although a far greater percentage operated in sectors where such TRIMs existed. In other words, TRIMs often fail to bind (UNCTC, 1991). However, these surveys did *not* take account of the firms that may have refrained from investing in countries with TRIMs. By discouraging FDI and distorting the allocation of global capital, the use of TRIMs by an individual country may have efficiency consequences for the world.

Whatever the economic rationale behind restrictive FDI policies, the available empirical evidence suggests that local content and related policies are costly to the economy. Furthermore, they often do not achieve the desired backward and forward linkages, and they encourage inefficient foreign entry and create potential problems for future liberalisation. Those who successfully enter a market when it is subject to some investment measures lobby against a change in regime (Moran, 1998). However, a case can be made that as long as trade barriers on final goods are low enough to allow firms to export, TRIMs may not be particularly costly (Hoekman and Saggi, 2000).

Finally, TRIMs are just part of the relevant policy landscape; investment policy measures that affect entry and operations are often general, not tied to trade performance or trade policy. Many countries impose licensing and related screening and approval regimes. These are often associated with related red tape costs for foreign investors. Some countries may also prohibit entry through FDI altogether, or impose equity limitations. Such policies may reflect welfare-enhancing attempts to shift foreign profits to the domestic economy or welfare-reducing rent-seeking activities by bureaucrats and special interest groups. The TRIMs agreement does not apply to such non-trade-related policies, nor does it affect service industries. The latter are covered by the GATS, however, which extends to FDI policies if countries make specific market access and national treatment commitments for this mode of supply.

5.2 The TRIPS Agreement

During the Uruguay round, multilateral negotiations on intellectual property rights were complicated both technically and politically (Hoekman and Kostecki, 2001). The resulting TRIPS agreement reflects the nature of these negotiations; a full discussion of the agreement is beyond the scope of this report. Here, we focus on the implications of the TRIPS agreement for technology transfer. In other words, the purpose of this report is not to dig into the rationales behind the positions of each side, but rather to ask whether or not the TRIPS agreement is expected to have a substantial effect on the nature of ITT in the post-Uruguay Round era.[20]

As one might expect, developing and developed countries see the TRIPS agreement on IPRs in very different ways. Supporters of the TRIPS agreement argue that stronger IPRs in developing countries would encourage R&D by developed country firms and that more R&D should lead to more innovations that would eventually spread to all parts of the world through trade, FDI and other channels. The opposing view is that stronger IPR enforcement could easily throttle local R&D in developing countries since most such R&D is geared toward adopting and adapting foreign technologies. If too stringent a standard is used, local firms in developing countries might be dissuaded from making investments in absorption of foreign technology. Maskus and McDaniel (1991) examine how the Japanese patent system affected post-war technical progress in Japan. As is well known, the Japanese patent system was designed to encourage incremental and adaptive innovation and diffusion of knowledge. They found that the Japanese system was quite successful in its goals and promoted the development of large numbers of utility model applications for incremental innovations that had a strong effect on productivity growth in Japan. It is far from obvious that the strategies adopted by Japan (and also Korea in some respects) would be feasible if the stipulations of the TRIPS agreement had been in force at that time.

The TRIPS agreement explicitly addresses the issue of ITT: Articles 66 and 67 of the TRIPs agreement commit developed countries to identify measures that can encourage technology transfer to developing countries and help improve the technological base of recipient countries. While the spirit behind these articles is laudable, they really cannot be taken seriously and seem at best to be lip service. After all, simply including such articles in the TRIPS agreement does nothing to alter the incentives of agents whose decision-making determines the costs and benefits of technology transfer.

Those opposed to the TRIPS agreement also claim that stronger IPR protection in developing countries will do little to encourage global innovation, while at the same time it will do great harm to social welfare in such countries by resulting in higher prices. In other words, the TRIPS agreement is not about economic efficiency but is rather about equity and rent sharing. It is difficult to argue against the point that a complete enforcement of IPRs by developing countries (as dictated by the TRIPS agreement) will result in large transfers from such countries to exporters of products that are intensive in the use of IPRs. McCalman (1999) finds that the implementation of TRIPS could result in a net transfer of over $8 billion (measured in 1995 US$) from developing to OECD countries. Thus, the transfers involved are not trivial and the TRIPS agreement can generate a win–win situation only if its effect on the rate of technological development is large enough to outweigh the static welfare costs it imposes on developing countries. No one can say with certainty whether this is so. Thus, a sceptical position on the TRIPS agreement seems reasonable at present.

6
Policy Lessons

What lessons can countries draw from the preceding analysis that can help them to formulate better policies? Below, we discuss what existing research tells us about some of the central policy questions.

- *Will increased international trade encourage economic growth?* The role of trade in encouraging growth hinges critically on the geographical scope (national versus international) of knowledge spillovers. Such knowledge spillovers are neither exclusively national nor international; they are probably both to some extent. However, spillovers are more likely to be national in scope for developing countries than for industrial ones. Consequently, whether R&D and high-technology production are carried out in close geographical proximity to such countries may indeed matter for their development. However, the evidence on this front is not strong enough to warrant the use of policies that try to ensure that particular activities in the production value chain (such as R&D) ought to be located locally for a country to benefit from them.

- *Does the 'type' of FDI matter?* There is no evidence to suggest that particular types of foreign investment are more valuable than others. One needs to be especially cautious of claims regarding large spillovers from certain hi-tech investments. In fact, investment in simple activities (such as transportation and other fundamental services) might be where large returns lie and such activities need not result in large technology spillovers. Industrialisation can be subject to large co-ordination failures and investment in local infrastructure can help resolve such failures, even though it might fail to result in technology spillovers to local firms. Thus, it would probably be counter-productive for developing countries to favour FDI into relatively 'hi-tech' activities at the expense of FDI in more mundane activities that help develop local infrastructure.

- *Are technology licensing and joint ventures preferable to FDI?* In the past, local policy often caused foreign firms to opt for licensing or joint ventures rather than FDI. But there is little or no empirical evidence to support the view that licensing or joint ventures are superior channels of technology transfer relative to FDI. To be fair, few careful studies have attempted to evaluate this question empirically. So the jury is still out on this issue. At a general level, it seems reasonable to expect greater involvement of local agents in foreign investment projects to increase the scope for technology transfer to the local economy. However, even FDI utilises local workers and managers and multinational firms are generally averse to licensing core technologies to local firms. Thus, a preference for local involvement in the form of licensing might compromise the quality of the technology transferred by foreign firms.

- ***Do large multinationals misuse their market power?*** Small developing countries might be worried about the market power that large multinationals can exercise with respect to their small local competitors. This is a valid concern and should not be treated lightly. However, the solution is not to turn protectionist since that further increases the market power available to multinationals already in the market. Another fruitless approach is to try to secure exclusive deals with foreign investors. In fact, exactly the opposite needs to be done: small developing countries need to ensure that there is unbridled competition between all potential foreign investors. Even if market concentration results from the nature of the underlying production technology (i.e. large fixed costs), potential competition can still serve as a disciplining device for incumbent firms. However, for such competition to matter, the market should not be closed via trade policy restrictions.

- ***Are investment incentives a good policy?*** Policies designed to lure in FDI have proliferated in recent years but it is difficult to make the case in favour of such policies on basis of positive spillovers from FDI to domestic firms. Several recent plant-level studies have failed to find positive spillovers from FDI to their direct competitors. However, these studies require careful interpretation because they treat FDI as exogenous. In addition, FDI spillovers may be vertical in nature rather than horizontal (as is assumed in such studies). In fact, recent empirical evidence is strongly supportive of the presence of vertical spillovers. As noted earlier, for a subsidy to be used, one needs to argue that multinationals invest too little in such activities. However, the evidence suggests that they actually make substantial investments in improving the quality of their local suppliers. Whether these investments are socially optimal is a difficult issue to resolve. The use of investment incentives and other such policies is especially problematic when such subsidies are given only to a few firms because then they suppress competition amongst investors. When used in this way, investment incentives can be doubly costly: the subsidy given imposes a direct financial cost as well an indirect cost by reducing competition.

- ***Are trade and investment liberalisation sufficient to encourage technology transfer?*** Several studies (both theoretical and empirical) indicate that absorptive capacity in the host country is crucial for obtaining significant benefits from FDI. Without adequate human capital or investments in R&D, spillovers from FDI may simply be infeasible. Thus, liberalisation of trade and FDI policies needs to be complemented by appropriate policy measures with respect to education, R&D and human capital accumulation if developing countries are to take full advantage of increased trade and FDI. In fact, domestic policies that improve absorptive capacity might be of higher order importance than openness to trade and investment. This point cannot be over-emphasised. In their rush to implement policies that protect local agents against the market power of multinationals, developing country governments may forget that local technological capability cannot be enhanced without adequate investments in local education and R&D.

- *How does the enforcement of IPRs affect technology transfer and trade?* Empirical evidence supports the argument that IPRs are trade related and that asymmetric IPR protection across countries distorts the pattern of world trade. Furthermore, a country's IPR policy can alter the composition of FDI both at the industry level as well as at the level of the firm. In industries in which IPRs are crucial, firms may refrain from FDI if IPR protection is weak in the host country, or they may not invest in manufacturing and R&D activities. Lastly, IPR policy can also lead foreign firms to choose FDI over other arms-length modes of technology transfer, such as licensing.

- *Should developing countries utilise TRIMs?* The two types of TRIMs that have received the most attention are domestic content requirements and export performance requirements. There are certainly theoretical models that can justify the use of such TRIMs on welfare grounds. However, models in which TRIMs can improve welfare do not yield prescriptions that are precise enough to be implemented in the real world. Thus, it would be easy to get the policy wrong. Empirical evidence on the efficacy of TRIMS is mixed. While few (if any) studies support the use of domestic content requirements, the same is not true of export performance requirements. For example, such policies have been found to be effective in South Korea and Mexico. However, even in such cases, it is not always clear that the policy actually forced firms to export – many firms would have done so even in the absence of any policy directives.

- *Is the TRIPS agreement useful for encouraging technology transfer?* This question is of central importance to this report and the answer here is not as clear-cut as is often suggested by proponents of the TRIPS agreement. While it is true that IPR protection can indeed have a significant effect on innovation incentives in certain industries, it is worth bearing in mind that the main purpose of IPR protection is to limit the use of proprietary technologies. In this sense, IPRs are antithetical to technology transfer, almost by definition. However, this not the whole story. If firms feel more secure about their IPRs in developing countries, they may be more willing to engage in market transactions such as technology licensing to firms in developing countries, thereby leading to more technology transfer. The sharp divide between developed and developing countries on the TRIPS agreement reflects the fundamental conflict between their goals: developed countries wish to earn rents on their innovations, whereas developing ones want to learn how to innovate by assimilating existing technologies at relatively low cost.

Table 1.1: Global Exports of Capital Goods as a Percentage of Total Exports (1975–1996)

1975	23.5	1986	27.9
1976	23.8	1987	28.3
1977	24.1	1988	28.5
1978	25.0	1989	28.0
1979	22.7	1990	28.4
1980	21.5	1991	28.6
1981	22.2	1992	27.8
1982	23.3	1993	28.8
1983	23.9	1994	30.2
1984	24.6	1995	30.6
1985	25.8	1996	30.7

Source: *International Trade Statistics Yearbook*, 1983–98

Table 1.2a: FDI Inward Stock, 1980–2000 ($ billion)

	1980	1985	1990	1995	2000
World	616	894	1889	2938	6314
% in Developed	60.9	61.1	74.1	69.8	66.7

Source: UNCTAD, 2001

Table 1.2b: FDI Inward Stock as a Percentage of GDP

	1980 (%)	1985 (%)	1990 (%)	1995 (%)	1999 %)
World	6.0	7.8	9.2	10.3	17.3
Developed Countries	4.7	6.1	8.4	9.2	14.5
Developing Countries	10.2	14.1	13.4	15.6	28.0

Source: UNCTAD, 2001

Figure 1.1: FDI Inflows as a Percentage of GDI (Low- and Middle-income Countries)

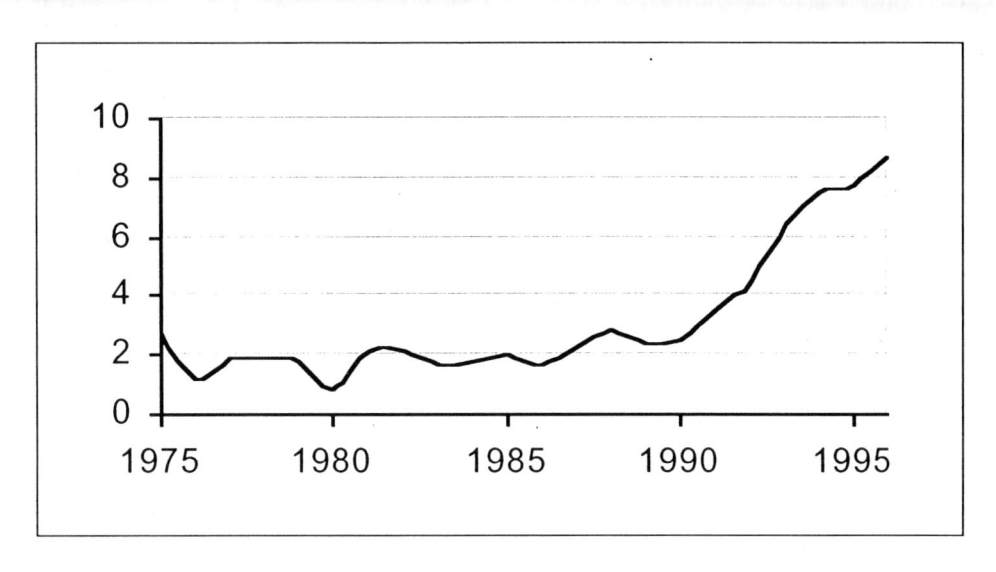

Table 1.3: Receipts of Royalties and Licence Fees, 1985–97 ($ million)

		Germany			United States	
	Total	Intra-firm		Total	Intra-firm	
		German parent firms only	Foreign affiliates in Germany		US parent firms only	Foreign affiliates in US
Year						
1985	546	464	83	6680	–	–
1986	780	597	122	8114	5994	180
1987	997	698	146	10183	7668	229
1988	1081	883	124	12147	9238	263
1989	1122	899	106	13818	10612	349
1990	1547	1210	235	16635	12867	383
1991	1515	–	345	17819	13523	583
1992	1680	–	472	20841	14925	733
1993	1596	–	498	21694	14936	752
1994	1720	–	489	26712	19250	1025
1995	2174	1486	642	30289	21399	1460
1996	2315	1667	653	32823	22781	1929
1997	2282	1659	509	33676	23457	2058

Source: UNCTAD, 1999.

Notes

1 A substantial literature supports the view that more open economies grow faster (see, for example, Dollar (1992) and Sachs and Werner (1995)). However, see Rodriguez and Rodrik (1999) for a critical analysis of this literature.
2 For the purposes of this report, openness means openness toward international trade and FDI.
3 Pack (1992) provides an overview of what can be reasonably expected in terms of technology transfer to developing countries, given that the potential for transfers is large.
4 A companion report entitled *Encouraging Technology Transfer to Developing Countries: Role of the WTO*, provides a more comprehensive discussion of the major policy implications of this report.
5 Of course, the maintenance of such monopoly power will often require protection via intellectual property rights.
6 Although R&D based endogenous growth theory is quite appealing theoretically, empirical evidence does not provide a strong endorsement (Pack, 1994). In fact, Jones (1995a and 1995b) shows that the data reject the empirical implications of the early variants of these models. The newer strand of growth theory has focused on creating models that do not have the 'scale effects' that lack empirical support.
7 In North–South models, the more interesting question is how Southern imitation affects incentives for innovation in the North.
8 In their micro-level study of the semiconductor industry, Irwin and Klenow (1994) find that learning (resulting from production) spills over as much across national borders as it does within.
9 However, in recent years developing countries have been getting a smaller share of global FDI flows: in 1999 and 2000, their share has been around 20 per cent (UNCTAD, 2001).
10 For the least developed countries, this ratio is about 14 per cent (UNCTAD, 2001).
11 Of course, a country may also gain access to foreign technology via outward FDI (see Lichtenberg and Potterie, 2001). However, developing countries do not undertake much outward FDI in countries with more advanced technologies than them. As a result, inward FDI is likely to be the main channel of technology transfer for developing countries.
12 It is conceivable, however, that, due to their larger size and other advantages that they enjoy over local firms, multinationals can alter the market outcome in their favour despite technology leakage. Assuming the rate of increase in efficiency of the local firms to be positively related to the scale of operation of the multinational firm's subsidiary, Das (1987) shows that despite technology leakage, a multinational may find it profitable to transfer technology.
13 In an oligopoly model, Saggi (2002) finds that the entry of a multinational firm generates net additional linkages only when the technological advantage of the multinational over its local competitor is moderate.
14 Much of the literature has been preoccupied with the effect of FDI on the technological capabilities of host firms and the industry, whereas FDI can also enhance productivity via its effect on management practices (see Child et al., 2000).
15 The authors also note that the large tax breaks and incentive packages given to multinationals seem out of proportion in relation to the magnitude of spillovers they generate.
16 Nadiri (1991) found that an increase in the US-owned capital stock has had a positive effect on the productivity growth of manufacturing industries in France, Germany, Japan and the UK. Thus, his results support the view that inward FDI generates positive spillovers in the host country.
17 Note, however, that export performance requirements are not covered by the TRIMs agreement.
18 However, this is not required. The standard example of a policy distortion is trade protection. Consider a developing country with protection on capital-intensive goods in a standard two sector, two factor model. Allowing in foreign capital causes the output mix to shift towards the capital-intensive sector, so that imports of the capital-intensive goods and, therefore, tariff revenues fall. This reduces welfare because each unit of imports is worth more inside the country than its cost from world markets (Hamada, 1974).
19 See Glass and Saggi (1999b) for an analysis of a situation where such policies have distributional consequences across countries.
20 It is too soon to say what effect the TRIPS agreement has already had on ITT.

II

Encouraging Technology Transfer to Developing Countries: The Role of the WTO

1
Introduction

International technology transfer is a complex phenomenon and its importance for economic development can hardly be overstated. Accordingly, Paragraph 37 of the WTO Doha Ministerial Declaration called for the establishment of a Working Group on Trade and Technology Transfer:

> We agree to an examination, in a Working Group under the auspices of the General Council, of the Relationship between Trade and the Transfer of Technology, and any of the possible recommendations that might be taken within the mandate of the World Trade Organization (WTO) to increase flows of Technology to developing countries. The General Council shall report to the Fifth Session of the Ministerial Conference on progress in the examination.

The Working Group on Trade and Technology Transfer (WGTT) has been in operation for over a year and has been involved in a variety of tasks. It has hosted presentations in trade and technology transfer by intergovernmental organisations, academia and by member countries regarding their own experiences; obtained background papers from the WTO Secretariat; discussed submissions from member countries; and publicised work being done by other WTO bodies on trade and ITT. Through these multi-faceted efforts, it is hoped that member countries can gain a better understanding of the process of ITT.

At its first informal meeting on 6 March 2002, the WGTT asked the WTO Secretariat to produce a factual survey of the issues related to trade and the transfer of technology that was intended to serve as a basis for discussion at its first formal meeting. The result of this request was a comprehensive document (WT/WGTT/W/1) which, together with Part I of this report, provides a thorough overview of the economics of international technology transfer.

Several delegations to the WTO have correctly noted that ITT is a broad phenomenon and that the work of the WGTT should include a variety of viewpoints (WT/WGTT/M/1). However, too inclusive a definition of ITT can easily thwart meaningful discussion. For example, ITT also occurs when scientists in developing countries read scientific literature produced by their counterparts in developed countries (and vice versa). However, this channel of ITT not only falls outside the scope of the private sector, but is also difficult to measure. Furthermore, it is difficult to see what, if anything, the WTO can do to encourage this process. Accordingly, this paper restricts attention to those aspects of ITT that are closely related to the mission of the WTO.

A major reason for focusing on trade and FDI as channels of ITT is that it is practical to do so; while policies should, in principle, be designed to take advantage of all existing channels of ITT, effective policy implementation and enforcement requires that policies are based on channels that are relatively easy to measure and quantify (which trade and FDI surely are).

While economic analysis provides us with a coherent framework for analyzing ITT, many aspects of ITT remain poorly understood. This situation does not necessarily reflect on the discipline of economics – the very nature of ITT makes analytical and empirical progress difficult. At the heart of ITT is the exchange of information and knowledge, both of which are difficult to quantify. In fact, market transactions in ITT are hampered by several problems of which the following three are crucial: asymmetric information; market power; and positive externalities.

1.1 Asymmetric Information

By definition, technology transfer involves an exchange of information between those who have it and those who do not. It is well known that the presence of asymmetric information can hamper the efficient working of markets. In simple words, the problem is the following: how does someone possessing useful information/technology convey to the uninformed party the value of information which the latter lacks? The informed party must convey the information in order for its value to be assessed by the uninformed party.

The problem, of course, is that as soon as the valuable information is shared, the basis for trade disappears since both parties know the same thing! This line of reasoning, no doubt, over-simplifies the issues, but it does capture something fundamental about the nature of technology as an economic good. In fact, it is widely acknowledged that the presence of asymmetric information can lead to large transactions costs that can stifle market-mediated technology transfer. In the international context, technology transfer faces additional hurdles: information problems are more severe and the enforcement of contracts more difficult to achieve. In fact, the received theory of the multinational firm holds that such firms establish subsidiaries in foreign markets because they find it difficult to use markets to profit from their proprietary technologies.

1.2 Market Power

Yet another serious issue confronting ITT is that owners of new technologies typically have substantial market power which often results from patents and other IPRs. Such market power necessarily implies that technology will not be transferred at marginal cost – i.e. its price will be higher than the socially optimal price (given that the technology exists). Of course, this divergence between price and cost is what allows innovators to profit from their innovation. Nevertheless, the presence of market power can in principle create room for government intervention.

1.3 Positive Externalities

The third important problem affecting ITT is the presence of externalities. Simply put, there are externalities from a transaction in technology if the costs and benefits of the exchange are not fully internalised by those involved in the exchange. For example, if the subsidiary of a multinational firm receives a new technology from which unrelated local agents (say, in some other industries) derive some benefit, the parent firm's decision regarding the extent of technology transfer will typically fail to account for the positive externalities enjoyed by those local agents. In fact, document WT/WGTT/W/3 of the WGTT also provides another useful example:

> Consider, for example, the case of a new technology that would lower fuel consumption, where the environmental benefits of lower fuel consumption were not taken into account by private investors in the new technology because these investors would not be able to capture fully the returns from environmental benefits to society as a whole. Without government intervention in this case, markets would fail to capture the full benefits of the new technology. More generally, whenever private and social costs or benefits do not coincide, a case might be made for government intervention. Many governments have pursued policies which reflect the conviction that without intervention, there will be underinvestment in socially beneficial technology.

It is important to bear in mind, however, that the focus of the WGTT is on ITT and not on technology creation. This might appear to be a semantic distinction but it is not. For there to be a rationale for active government policy in the area of ITT on the basis of externalities, there need to be externalities in the process of transfer of technology. Of course, as noted above, there are good reasons to believe that this is so.

Due to the problems discussed above, there is potential for governments to play an important role in the process of international ITT. However, government policy can only be effective if it alters the incentives of private agents that possess innovative technologies. Furthermore, the cost of such government intervention should not be ignored. The potential for welfare-improving government policy does not always transfer into the implementation of such policies. The domestic regulatory and political environment of a country needs to prevent policies from becoming hostage to rent-seeking and lobbying activities. Otherwise, well-intentioned policies can do more harm than good.

2
Trade as a Channel of Technology Transfer

Traditional economic theory argues that expansion in world trade yields efficiency gains by improving the global allocation of resources. The more interesting question is whether trade also yields dynamic efficiency gains by improving productivity growth in the world. Considerable economic research has addressed to this question and its main findings are discussed below.

2.1 Research on Growth Effects of Trade

Standard neoclassical growth models assume costless ITT by positing a common production function across countries. However, this is an unsatisfactory approach – it is unlikely that firms in different countries can access the global pool of technologies at the same cost. In fact, recent research has shown that barriers to technology adoption are a key determinant of international differences in per capita income. Increased openness to trade can increase economic growth by lowering the barriers to technology adoption.

Recent endogenous growth models assign a central role to technological change and human capital accumulation. These models formalise the Schumpeterian notion of 'creative destruction' and are built on the idea that entrepreneurs conduct R&D to profit from the monopoly power that results from innovation. A discussion of the main assumptions underlying these growth models is useful for shedding light on the relationship between trade and ITT. In one strand of such models, growth is sustained through the assumption that the creation of new products expands the knowledge stock, which then lowers the cost of innovation. As more products are invented, both the costs of inventing new products and the profits of subsequent innovators are lower because of increased competition (no products disappear from the market in this model). A second strand of growth models is built on the idea that consumers are willing to pay a premium for higher quality products. As a result, firms always have an incentive to improve the quality of products. The important assumption that sustains growth in the quality-based model is that every successful innovation allows all firms to study the attributes of the newly invented product and then improve on it. Patent rights restrict a firm from producing a product invented by some other firm, but not from using the knowledge (created by R&D) that is embodied in that product. Thus, as soon as a product is created, the knowledge needed for its production becomes available to all; such knowledge spillovers ensure that anyone can try to invent a higher quality version of the same product. Both types of growth models contain an important insight: since new products result from new ideas, trade in goods can help transmit embodied knowledge internationally.

Multi-country versions of endogenous growth models have two strands: those that study trade between identical countries and those that have a North–South structure. Although knowledge spillovers are central to both, technology transfer is a central feature only of North–South models. North–South models that emphasise the product-cycle nature of trade have been particularly useful for understanding ITT and merit some further discussion. These product-cycle models assume that new products are invented in the North and that, due to the lower relative Southern wage, Southern firms can successfully undercut Northern producers by imitating Northern products. A typical good is initially produced in the North until either further innovation or successful Southern imitation makes profitable production in the North unprofitable. Consequently, production either ceases (due to innovation) or shifts to the South (due to imitation). Thus, prior to imitation, all products are exported by the North, whereas post-imitation they are imported, thereby completing the cycle. These models capture technology-driven trade and have been generalised to consider technology transfer more explicitly. Neither FDI nor licensing (choices available to innovators for producing in the South) was considered in the early variants of these models.

What do endogenous growth models imply about the effect of trade on productivity and growth? An important conclusion of this line of research is that much of importance hinges on whether knowledge spillovers are national or international in nature. If knowledge spillovers are international, these models endorse the view that trade is an engine of growth. However, when knowledge spillovers are national in scope, perverse possibilities can arise. Note that this perspective is more relevant for North–North models of trade because international knowledge spillovers (of one form or another) are assumed in North–South product-cycle models of trade, where the South is modelled as a pure imitator. Unfortunately, empirical evidence regarding the scope of knowledge spillovers has been somewhat mixed in nature. A broad generalisation of this evidence is that more studies argue in favour of international, rather than national, knowledge spillovers.

2.2 Capital Goods Trade

In principle, trade in both consumption and capital goods can contribute to technology transfer and the empirical studies discussed above typically utilise a country's imports of all goods while attempting to measure knowledge spillovers through trade. For example, when a country imports a manufactured consumption good (such as an automobile) local firms can absorb some technological know-how by simply studying the design and the engine of the imported automobile. While such attempts at reverse engineering are no doubt important, they probably contribute less to technology transfer than trade in capital goods (such as machinery and equipment) that are used in the production of other consumption goods. Trade in capital goods is more relevant than total trade for measuring knowledge spillovers because capital goods have higher technological content than consumption goods. In fact, it has been shown that variation in capital goods trade can better explain cross-country variation in productivity than can overall trade.

Overall, capital goods trade is a prominent part of world trade and its importance has increased over time. In 1975, while approximately 23 per cent of total trade in the world was trade in capital goods (defined as machinery and transport equipment); this ratio was over 30 per cent in 1996. During the period 1975–96, worldwide exports of capital goods as a percentage of GDP increased from about 4.2 per cent to approximately 7 per cent. In 1996, approximately 30 per cent of capital goods exports were destined for developing countries. Although the developing country share of imports of capital goods has increased over time, this increase has not been substantial (it was 28.9 per cent in 1980 compared with 30.8 per cent in 1996). Furthermore, developing countries are heavy importers of capital goods and some 85 per cent of the imports of machinery and transport equipment into developing countries come from developed countries.

3
The Role of Foreign Direct Investment

Intra-firm trade, that is trade between subsidiaries and headquarters of multinational firms, may account for one-third of total world trade. The importance of FDI can also be gauged from the fact that sales of subsidiaries of multinational firms now exceed worldwide exports of goods and services. Thus, FDI is the dominant channel through which firms serve customers in foreign markets.

Much of the flows of FDI occur between industrial countries (as does most intra-industry trade). For example, during 1987–92, industrial countries attracted $137 billion of FDI inflows a year on average; developing countries attracted only $35 billion, or slightly more than 20 per cent of global FDI inflows. Yet developing countries are becoming increasingly important host countries for FDI, especially because of substantial economic liberalisation undertaken by countries such as China and, to a lesser extent, India. During 1996 and 1997, over 40 per cent of global FDI flows went to developing countries. The recent surge in capital flows to developing countries, of which FDI has been a significant part, is also reflected in the fact that approximately 32 per cent of the total stock of FDI is now in developing countries. Because of their smaller size, FDI is of relatively greater importance to developing countries. In 1999, the total inward stock of FDI as a percentage of GDP was almost 28 per cent in developing countries, compared with less than 15 per cent in industrial countries.

3.1 Multinational Firms and Technology Transfer

Multinationals are creatures of market imperfection; they arise because markets are not always the most efficient means of exploiting intangible assets (such as technology) in international markets. Almost invariably, the very assets that lie behind the emergence of multinationals also create market power. Thus, while writing formal models of multinational firms, the widely prevalent model of perfect competition must be abandoned. A well-accepted stylised fact is that the presence of multinationals is positively correlated with market concentration. Developing countries are often worried about the market power that large multinationals can exercise with respect to their small local competitors. This is a valid concern and should not be taken lightly because many such countries do not have the resources to enforce competition law. Of course, this is not to say that restricting the entry of multinationals would make markets more competitive in developing countries. On the contrary, the restrictions faced by multinationals probably worsen market concentration in many international markets.

Multinational firms play a crucial role in ITT. For example, in 1995 over 80 per cent of global royalty payments for international transfers of technology were made from subsidiaries to their parent firms. In general, technology payments and receipts have risen steadily over time, reflecting the importance of technology in international production. The data also indicate the importance of FDI for international trade in technology. During 1985–97, between two-thirds and nine-tenths of technology flows were intra-firm in nature. Furthermore, the intra-firm share of technology flows has increased over time. Royalty payments only record the explicit sale of technology and do not capture the full magnitude of technology transfer through FDI relative to technology transfer via imitation, trade in goods and other channels.

3.2 Spillovers from Foreign Direct Investment

By encouraging inward FDI, developing countries hope not only to import more efficient foreign technologies, but also to improve the productivity of local firms via technological spillovers to them. The central difficulty in determining whether this actually happens or not is that spillovers do not leave a paper trail; they are externalities that the market fails to take into account. Nevertheless, several studies have attempted the difficult task of quantifying spillovers. Conceptually, spillovers can arise via the following channels: (i) demonstration effects – local firms may adopt technologies introduced by multinational firms through imitation or reverse engineering; (ii) labour turnover – workers trained or previously employed by the multinational may transfer important information to local firms by switching employers, or may contribute to technology diffusion by starting their own firms; and (iii) vertical linkages – multinationals may transfer technology to firms that are potential suppliers of intermediate goods or buyers of their own products.

Early efforts in search of horizontal spillovers from FDI proceeded by relating the inter-industry variation in productivity to the extent of FDI. By and large, these studies find that sectors with a higher level of foreign involvement (as measured by the share of the labour force in the industry employed by foreign firms or the extent of foreign ownership) tend to have higher productivity, higher productivity growth or both. Of course, correlation is not causation and this literature may overstate the positive impact of FDI on local productivity. Investment may have been attracted to the more productive sectors of the economy instead of being the cause of the high productivity in such sectors. In other words, the studies ignore an important self-selection problem. Only plant-level studies can control for the self-selection problem that may plague industry-level studies. Several plant level studies have found that there are no positive spillovers from subsidiaries of multinationals to their domestic rivals. In fact, some even find that productivity in domestic plants declines with an increase in foreign investment. On the other hand, some recent plant level studies find that there are positive spillovers from FDI at the industry level. It is worth emphasising that even if studies fail to find positive spillovers from FDI, it does not imply that host countries have nothing significant to gain (or that they must lose) from FDI. Domestic firms should be expected to suffer from an increase in competition; in fact, part of the benefit of inward FDI is that it can help weed out relatively inefficient domestic firms. Resources released in this process will be put to better use by foreign firms with superior technologies, efficient new entrants (both domestic and foreign) or some other sectors of the economy. However, such reallocation of resources cannot take place instantaneously. Existing studies of spillovers do not cover a long enough period to be able to accurately determine how FDI affects turnover rates (entry and exit). Furthermore, by design, such horizontal studies cannot identify linkages and spillovers that may result from FDI in industries other than the one in which FDI occurs. Finally, multinationals may benefit host countries even if they fail to generate technology spillovers for their local competitors.

The preceding discussion suggests that spillovers to local firms that directly compete with the multinationals would be the most elusive of benefits that host countries can expect to enjoy from FDI. Local agents other than domestic competitors of multinationals (for example local workers and local suppliers) are more likely to enjoy positive externalities from FDI. Spillovers may also be of an entirely different nature than the ones explored in plant-level studies of productivity: local firms may enjoy positive externalities that make it easier for them to export to foreign markets. Such externalities may come about because better infrastructure (transportation, storage facilities and ports) emerges in regions with a high concentration of foreign exporters.

Although researchers have extensively studied imitation and reverse engineering as channels of inter-firm technology diffusion, they have tended to neglect the role of labour turnover. Labour turnover differs from the other channels because knowledge embodied in the labour force moves across firms only through the physical movement of workers. Whatever little evidence that there is on labour turnover is itself of a mixed nature. While some studies have found that labour turnover from multinationals to local firms is rather scarce, others have documented the opposite. One possible interpretation of the existing evidence is that in countries such as South Korea and Taiwan, where local firms are not too far behind multinationals (unlike local firms in many African economies), labour turnover is more likely to happen. In other words, the ability of local firms to absorb technologies introduced by multinationals may be a key determinant of whether labour turnover occurs as a means of technology diffusion in equilibrium. Furthermore, the local investment climate may be such that workers who wish to leave multinationals in search of new opportunities (or other local entrepreneurs) find it unprofitable to start their own companies, implying that the only alternative opportunity is to join existing local firms. The presence of weak local competitors probably goes hand in hand with the lack of entrepreneurial efforts because both may result from the underlying structure of the economic environment.

Vertical technology transfer from buyers to sellers has been documented to occur when firms from industrialised countries chose to buy the output of firms in many Asian economies in order to sell it under their own name. For example, companies such as Radio Shack and Texas Instruments have commissioned firms in developing countries to produce components or entire products, which are then sold under the retailer's name. The relations between Korean firms and the foreign buyers went far beyond the negotiation and fulfillment of contracts. Almost half of the firms said they had directly benefited from the technical information foreign buyers provided. The knowledge transfers involved were multi-faceted: not only manufacturing knowledge was transferred but exact sizes, colours, labels, packing materials and instructions to users. It has also been found that in the later 1970s, many importing firms from industrialised countries maintained very large staffs in countries such as Korea and Taiwan who spent considerable time with their local manufacturers assisting them in meeting their specifications.

More recent evidence regarding vertical technology transfer is provided by Mexico's experience with the *maquildora* sector and its automobile industry. Mexico started the *maquiladora* sector as part of its Border Industrialisation Programme designed to attract foreign manufacturing facilities along the US-Mexico border. Most *maquiladoras* began as subsidiaries of US firms that shifted labour-intensive assembly operations to Mexico because of its low wages relative to the US. However, the industry evolved over time and the *maquiladoras* now employ sophisticated production techniques, many of which have been imported from the US. In the automobile industry, one of Mexico's most dynamic, FDI has resulted in extensive backward linkages: within five years of initial investments by US firms, there were hundreds of domestic producers of parts and accessories. US firms and other multinational firms transferred technology to these Mexican suppliers; industry best practices, zero defect procedures, production audits+ etc. were introduced to domestic suppliers, thereby improving their productivity. As a result of increased competition and efficiency, Mexican exports in the automobile industry have boomed.

Several developing country delegations have expressed concern about encouraging technological development that leads to higher exports (see WT/WGTT/M/4). Developing country firms face several barriers when trying to export to the rest of the world. They may not only lack the technology to produce high-quality goods but also not be fully aware of consumer tastes and preferences in other markets. The above discussion of Mexico's experience with FDI shows that both of these problems can potentially be solved via international outsourcing and the development of backward linkages (see WT/WGTT/M/4).

4
National Policies

This section discusses the implications for national policies with respect to trade, FDI and technology transfer that can be drawn from existing research. The literature that investigates the effect of trade protection on technology transfer and growth is too voluminous to receive adequate treatment here. As can be expected from models in which increasing returns, imperfect competition and externalities play a central role, the results depend on the details of a particular model and require careful interpretation. To the extent that one can draw a general conclusion from such a complex literature, it is that this literature does not provide an unconditional argument against trade protection. The conclusions hinge dramatically on the scope of knowledge spillovers: international knowledge spillovers strongly tilt the balance in favour of free trade, whereas national spillovers create a role for policy intervention that can combat path dependence resulting from a historical accident. For example, if productivity improvements depend only on a country's own R&D, a case can be made for policies that ensure that industries in which such improvements occur at a rapid rate are not all located elsewhere.

4.1 Policies Toward Foreign Direct Investment

There is no simple way of describing the policy environment that faces multinationals in developing countries. A roughly accurate statement is that while FDI in services markets faces a multitude of restrictions, FDI into the manufacturing sector is confronted with both restrictions and incentives, sometime in the same country. In countries that historically emphasised import-substituting industrialisation – such as most of Africa, Latin America and south-east Asia – FDI was either completely prohibited or multinational firms had to operate under severe restrictions. In fact, even in countries where technology acquisition was a major policy objective, multinationals were rarely permitted to operate fully-owned subsidiaries; Japan, Korea and Taiwan all imposed restrictions on FDI at various times. In other words, 'outward-oriented' economies were not particularly keen on allowing multinational firms into their markets.

Japan's Ministry of International Trade and Investment played an active role in the country's acquisition of foreign technology. MITI limited competition between potential Japanese buyers, did not allow inward FDI until 1970, never greatly liberalised FDI and even insisted at times that foreign firms share their technology with local firms as a precondition for doing business in Japan. By prohibiting FDI and placing other restrictions on the conduct of multinationals, government policies in many countries may have effectively weakened the bargaining position of foreign firms. In fact, in Japan, MITI restricted many local firms from participating as potential buyers exactly for this reason. Restrictions on FDI were often accompanied by a more lax attitude toward other modes of ITT. To fully understand the effect of such restrictions it is important to evaluate the broad FDI policy environment in the 'Japan-Korea' model. In contrast to their restrictive policies toward FDI, both Japan and South Korea aggressively encouraged licensing of foreign technology. The experience of South Korea and Japan is fairly representative in one important respect: there are few, if any, examples of countries that encouraged FDI but restricted technology licensing.

Sometimes FDI policy favours joint ventures relative to fully-owned subsidiaries of multinationals. For example, the Chinese government has been particularly interventionist in technology transactions and has encouraged FDI in the form of joint ventures. Although fully-owned subsidiaries of foreign firms are not prohibited from doing business in China, the policy environment favours joint ventures over such enterprises. Of course, an obvious reason for this might be that all such policies simply reflect protectionism. Large public firms or hitherto protected private firms who are unable to compete with multinationals may be able to secure protection through the political process. A more benevolent interpretation is that such policies seek to maximise technology transfer to local agents while limiting the rent erosion that results from the entry of multinational firms. Empirical evidence on this issue is scarce. But to the extent that it exists, it shows that the degree of foreign ownership affects neither the productivity of firms that get foreign equity nor the extent of spillovers to the domestic sector.

The focus on technology spillovers to local firms can also lead one to ignore a crucial issue: restrictions on FDI relative to licensing also need to account for the incentives of foreign firms regarding technology transfer under the two modes. Here, the evidence is less supportive of the Japan-Korea model. Several earlier studies document that technologies transferred to wholly-owned subsidiaries are of a newer vintage than licensed technologies or those transferred to joint ventures. Thus, by forcing multinationals to license their technologies, host countries might also be lowering the quality of technologies they receive.

Many countries still do not allow free entry of multinational firms and often express preferences with regard to the type of FDI: that is, entry by Pepsi or Coke is viewed differently from entry by GM or Texas Instruments. Unfortunately, the literature provides little insight for understanding such policies. Other than the standard argument that certain industries are able to secure greater protection than others, perhaps spillovers to the local economy are higher under certain types of FDI.

There is basically no evidence that suggests that particular types of foreign investment are more valuable than others. One needs to be especially cautious of claims regarding large spillovers from certain hi-tech investments. In fact, investment in simple activities (such as transportation and other fundamental services) might be where large returns lie and such activities need not result in large technology spillovers. Industrialisation can be subject to large co-ordination failures and investment in local infrastructure can help resolve such failures even though it might fail to result in technology spillovers to local firms. Thus, it would probably be counter-productive for developing countries to favour FDI into relatively 'hi-tech' activities at the expense of FDI in more mundane activities that help develop local infrastructure.

Developing countries have expressed the concern that they only receive obsolete technologies relative to those transferred to developed countries (see WT/WGTT/M/2). Unfortunately, there is little direct evidence on this issue. One way of confirming this claim is to measure the productivity of subsidiaries of multinational firms in different countries. More specifically, suppose a multinational firm has subsidiaries in two different markets: one developing (country A) and one developed (country B), where the two subsidiaries are involved in producing the same range of products. Then, if one finds that the plant in country A has much worse technology than the plant in country B, the above hypothesis can be confirmed. One reason such evidence simply does not exist is that most horizontal FDI (i.e. between two plants that make the same product in different countries) takes place among industrialised countries. Only developing countries with large domestic markets or with high levels of protection (that make exporting to the local market difficult) are likely to obtain horizontal FDI. Even if one could collect the type of evidence that is required to settle this issue, it is worth keeping in mind that market conditions, infrastructure and the business environment in developing countries might be better served by technologies that are modern relative to theirs, but that are not quite state of the art. In fact, private companies making decisions on ITT will optimise against market conditions and choose technologies that are appropriate from the viewpoint of profit maximisation. There is no guarantee that they will find it optimal to transfer state of the art technologies to developing countries.

Several delegations to the WTO have asked whether technology transfer should be a precondition for FDI (see WT/WGTT/M/2). The evidence discussed in this report shows that technology transfer requirements imposed by a host country will frequently be irrelevant since firms would undertake much technology transfer even in their absence. Yet, the magnitude of ITT undertaken by multinationals need not be socially optimal from the viewpoint of developing countries. The possibility of technology spillovers, together with the fact that multinationals possess significant market power, can easily result in too little ITT to the host country. However, for technology transfer measures to be effective, they would need to complement the incentives of multinational firms. For example, considerable evidence exists that multinationals are keen to transfer technology to their local suppliers. Policies that facilitate this process, as opposed to those that insist on ITT to local competitors of multinationals, are more likely to succeed.

The most frequently observed policy restrictions on FDI are in services markets; these have to do with the number of foreign firms permitted to enter the market as well as the extent of foreign ownership permitted. The pattern of these restrictions differs across countries and often across industries within countries. For instance, consider policy in basic telecommunications services. At one end, in the Philippines, a high degree of competition co-exists with limitations on foreign equity partnership. Bangladesh and Hong Kong are examples of countries that have no limitations on foreign ownership, but both have monopolies in the international telephony and oligopolies in other segments of the market. Pakistan and Sri Lanka have allowed limited foreign equity participation in monopolies to strategic investors, and deferred the introduction of competition for several years. Korea, however, is allowing increased foreign equity participation more gradually than competition. There is little rigorous research that can shed light on the effects of such restrictions. The only available insight is that certain types of policy restrictions on FDI might arise from attempts by local governments to improve local welfare in an environment of imperfect competition and costly technology transfer.

Despite the prevalence of various types of restrictions on FDI, multinationals do not face an entirely hostile environment in developing countries. In fact, many countries try to lure in large multinational firms via the use of investment incentives. Interestingly enough, it is not unusual to find investment incentives being offered in conjunction with performance requirements and other restrictions on FDI, perhaps to partially offset the negative impact of the latter on the likelihood of investment by multinationals. The schizophrenic nature of the overall policy environment reflects the guarded optimism with which many developing countries view the entry of multinational firms into their territory.

Perceptions about multinational firms and their effects on host countries have undergone a transformation in the last 50 or so years. Many countries that were quite opposed to permitting inward FDI now appear eager to attract FDI (particularly in the manufacturing sector). The clear evidence of this change in attitude is the proliferation of fiscal and financial incentives to foreign investors in the form of up-front subsidies, tax holidays and other grants. Basic economic theory tells us that it is optimal to subsidise an activity if it generates positive externalities – i.e. the activity benefits agents other than those directly involved in the activity itself. As is well known, activities that generate positive externalities tend to be under-provided by the market. Thus, market forces will generate too little FDI if such investment is accompanied by positive externalities.

Externalities could exist either between investors or from investors to other agents in the host country. However, if the dominant externalities are those that exist between potential investors, then it is unclear that host countries ought to be subsidising inward FDI. However, investment incentives may be justified if host countries enjoy externalities from inward FDI (such as technology spillovers for local firms). But the real issue is whether local suppliers enjoy benefits over and beyond the level that is perceived by the multinationals. For example, if a multinational firm helps a local supplier lower its costs by 10 per cent, and in turn enjoys a 10 per cent reduction in the price of the input it purchases from the supplier, then the multinational fully realises the benefit accruing to the supplier and there would be no positive spillovers. Of course, if the local supplier can secure some of the rents that come about from the cost reduction, the local economy will enjoy a positive externality that can potentially justify the use of investment incentives. Alternatively, if the process of cost reduction spills over from one local supplier to another, there can again be grounds for the use of investment incentives.

There is a long tradition in the management literature arguing the prevalence of 'follow-my-leader' approaches among multinational firms. The prevalence of such behaviour amongst multinationals suggests that an alternative case for the use of FDI incentives can be made on the basis of the oligopolistic nature of the markets within which FDI occurs. For example, consider Mexico's recent experience with FDI in its automobile industry. Initial investments by US car manufactures into Mexico were followed by investments not only by by Japanese and European car manufacturers, but also by firms who made automobile parts and components (i.e. their suppliers). As a result, competition in the automobile industry increased at multiple stages of production, thereby improving efficiency. Such a pattern of FDI behaviour (i.e. investment by one firm followed by investment by others) reflects strategic considerations involved in FDI decisions. Since multinational firms compete in highly concentrated markets, they are highly responsive to one another's decisions. An implication of this interdependence between competing multinationals is that a host country may be able to unleash a sequence of investments by successfully inducing FDI from one or two major firms. More broadly, if the local economy lacks a well-developed network of potential suppliers, multinational firms might be hesitant to invest and local suppliers may not develop because of lack of demand. In the presence of such interdependence, the development of an economy can be subject to a co-ordination problem that can be partially resolved by initiating investments from key firms. Of course, co-ordination problems in industrial development are too big an issue to be tackled only by the use of investment incentives, but such policies can certainly help.

Common sense suggests that if any set of policies should affect ITT and FDI, it will be the host country's intellectual property rights regime. Surveys of US multinational firms frequently find that such firms are more willing to invest in countries with stronger IPR protection Empirical evidence also indicates that the level of IPR protection in a country also affects the composition of FDI in two different ways. First, in industries for which IPRs are crucial (pharmaceuticals, for example), firms may refrain from investing in countries with weak IPR protection and instead export to such markets from alternative locations. Second, regardless of the industry in question, multinationals are less likely to set up manufacturing and R&D facilities in countries with weak IPR regimes, and more likely to set up sales and marketing ventures because the latter run no risk of technology leakage.

Research on the effects of IPR protection on economic growth indicates that there is a strong positive link between the two. It is also known that the effect of IPR protection is stronger for relatively open economies than it is for relatively closed ones. In other words, a strengthening of IPR protection is more conducive to growth when it is accompanied by a liberal trade policy. A possible interpretation of this finding is that by increasing foreign competition, trade liberalisation not only curtails the monopoly power granted by IPRs, but also ensures that such monopoly power is obtained only if the innovation is truly global. If firms in other countries can export freely to the domestic market and have better products or technologies, a domestic patent is useless in granting monopoly power.

To summarise, several studies (both theoretical and empirical) indicate that absorptive capacity in the host country is crucial for obtaining significant benefits from FDI. Without adequate human capital or investments in R&D, spillovers from FDI may simply be infeasible. Thus, liberalisation of trade and FDI policies needs to be complemented by appropriate policy measures with respect to education, R&D and human capital accumulation if developing countries are to take full advantage of increased trade and FDI. In fact, domestic policies that improve absorptive capacity might be of higher order importance than openness to trade and investment. This point cannot be over-emphasised. In their rush to implement policies that protect local agents against the market power of multinationals, developing country governments may forget that local technological capability cannot be enhanced without adequate investment in local education and human capital accumulation. As noted by several delegations in the fourth session of the WGTT, an open trading and investment system coupled with domestic efforts is the best means to enhance ITT in the long run.

5
Multilateral Rules and Disciplines

While there is no multilateral agreement that deals explicitly with technology transfer, several multilateral agreements of the WTO have direct bearing on ITT. In so far as the GATT (or for that matter the GATS) is concerned, its major objective of encouraging trade liberalisation can have only a positive effect on technology transfer. As was discussed earlier in this report, trade in goods (especially capital goods) plays a central role in diffusing technology internationally. Since there is nothing really controversial regarding the effect of GATT rules on technology transfer, the following discussion focuses on the Agreement on Trade-Related Intellectual Property Rights (TRIPS) and on the Agreement on Trade-Related Investment Measures (TRIMS). The TRIMS agreement is important because of the strong connection between technology transfer and FDI, whereas the TRIPS agreement deals explicitly with intellectual property rights.

5.1 TRIPS and ITT

In the preamble to the TRIPS agreement there is explicit recognition that the needs and concerns of the least developed countries do not coincide with those of the developed countries, particularly with respect to the protection of intellectual property rights. However, the agreement is silent on the details: what makes the needs of the least-developed countries special? How are they going to be offered flexibility, given that they are required to implement TRIPS (although they have been given more time to achieve compliance)? Furthermore, Article 7 of the TRIPS agreement notes that:

> *The protection and enforcement of intellectual property rights should contribute to the promotion of technological innovation and to the transfer and dissemination of technology, to the mutual advantage of producers and users of technological knowledge and in a manner conducive to social and economic welfare, and to a balance of rights and obligations.*

While the spirit behind this Article is commendable, it is far from clear how exactly a 'balance of rights and obligations' is to be achieved. Article 8 of TRIPS states that:

> *Appropriate measures, provided that they are consistent with the provisions of this Agreement, may be needed to prevent the abuse of intellectual property rights by right holders or the resort to practices which unreasonably restrain trade or adversely affect the international transfer of technology.*

Once again, it is not really clear what measures are deemed appropriate and what practices are viewed to be detrimental to ITT. For example, if a developing country views exclusive technology licensing as a trade restraint and/or a hindrance to technology transfer relative to multiple licensing, can it prevent a foreign firm from engaging in exclusive licensing? Would such a stance not violate the property rights of the firm? A communication by several developing countries (WT/WGTTT/W/6) notes that multinationals are reluctant to transfer technology through licensing because they do not want to create potential competition for themselves. This is certainly a valid argument and, as was noted earlier, many countries have prohibited FDI in order to indirectly encourage foreign firms to license their technologies. However, the use of such policies today is likely to be challenged by developed countries. A better strategy for developing countries would be to work with multinational firms and encourage the development of local suppliers which can then lead to future industrial development.

Article 66 commits developed countries to identify measures that can encourage ITT to developing countries and to help improve the technological base of recipient countries:

> *Developed country Members shall provide incentives to enterprises and institutions in their territories for the purpose of promoting and encouraging ITT to least-developed country Members in order to enable them to create a sound and viable technological base.*

By itself, this article does nothing to alter the incentives of agents whose decision-making determines the costs and benefits of ITT. Developed countries do not appear to have taken additional measures to promote ITT – the discussion of existing policies submitted by several developed countries to the WGTT suggests that such policies were already in place prior to TRIPS. Since TRIPS requires additional obligations on the part of the developing countries, it appears that they would be more willing to undertake those obligations if developed countries stepped up their efforts at promoting ITT. In fact, it seems reasonable to argue that developed countries might consider providing tax breaks and other financial incentives to companies that are involved in transferring technology to developing countries. It is also worth noting that Article 66 applies only to least-developed countries whereas it ought to apply to all developing countries. However, local absorptive capacity is crucial for successful technology transfer and such capacity is especially lacking in the least developed countries. Thus, in some senses, the explicit goal of encouraging ITT to only the least developed countries may give developed countries an easy excuse for the lack of substantive results.

Developing countries have noted that there should be greater repatriation of human resources from developed to developing countries to assist in the technological development of the latter (see WT/WGTT/M/2). Many individuals partly trained in developing countries have made important contributions to technological advances in industrialised countries (consider the role of Asian immigrants in the development of

Silicon Valley in the United States). Such immigrants could no doubt make an important contribution to the development of their home countries. However, they need an enabling environment and adequate infrastructure to be successful in their endeavours. National policies that make local conditions more attractive to such individuals would automatically encourage them to return to their countries of origin.

Article 66.2 of the TRIPS agreement can be implemented more effectively if developed country governments (especially those of the EU and the US) encourage their educational institutions to recruit and train students from less developed countries. Special programmes can and should be set up in science and engineering departments where students from less developed countries can be educated and trained at very low cost with the condition that they return to their countries for significant periods of time. While such programmes already exist in many countries, their scale is too small to alter the fundamental problems confronting many developing countries that need to absorb foreign technologies on a sustained basis before they can consider being involved in global innovation. To achieve this task, they need access to a supply of students and personnel that have exposure to advanced technologies and have the ability to tap the global resource pool on a continuous basis.

Perhaps the biggest issue of all regarding the TRIPS agreement is whether its implementation will lead to more technology transfer to developing countries (see WT/WGTT/M/2). The answer here is not as clear-cut as is often suggested by proponents of the agreement. While it its true that IPR protection can indeed have a significant effect on innovation incentives in certain industries, it is worth bearing in mind that the main purpose of IPR protection is to limit the unauthorised use of proprietary technologies. In this sense, IPRs are antithetical to ITT, almost by definition. However, this not the whole story. If firms feel more secure about their IPRs in developing countries, they may be more willing to engage in market transactions such as technology licensing to firms in developing countries, thereby leading to more ITT.

Those opposed to the TRIPS agreement also claim that stronger IPR protection in developing countries will not do much to encourage global innovation, while at the same time it will do great harm to social welfare in such countries by resulting in higher prices. In other words, the TRIPS agreement is not about economic efficiency, but rather about equity and rent sharing. It is difficult to argue against the point that a complete enforcement of IPRs by developing countries (as dictated by the TRIPS agreement) will result in large transfers from such countries to exporters of products that are intensive in the use of IPRs. In fact, recent research has shown that the implementation of TRIPS could result in large transfers from developing to developed countries. According to some calculations, the net transfer from developing to OECD countries would be in excess of $8 billion (measured in 1995 US$). Thus, the transfers involved are not trivial and the TRIPS agreement can generate a win–win situation only if its effect on the rate of technological development is large enough to outweigh the static welfare costs it imposes on developing countries. No one can say with certainty whether such is the case or not. Thus, a sceptical position on the TRIPS agreement seems reasonable at present.

The sharp divide between developed and developing countries on the TRIPS agreement reflects the fundamental conflict between their goals: developed countries wish to earn rents on their innovations, whereas developing ones want to learn how to innovate by assimilating existing technologies at relatively low cost. This basic tension cannot be just wished away and enforcing IPRs symmetrically across the world at the level at which they are enforced by industrialised countries does not seem to be entirely in the interest of developing countries. Furthermore, as is noted in the document WT/WGTT/W/6:

> *It is well known that the ongoing process of globalisation is rather skewed. While barriers to investment are coming down rapidly and consequently capital is becoming highly mobile, the mobility of other factors of production like labour and technology is becoming increasingly restricted ... It cannot be anybody's case that only those topics/subjects/issues where developing countries have to undertake commitments without receiving commensurate benefits, should be brought into the WTO.*

In other words, there is a fundamental asymmetry between the mobility of capital and labour in today's global economy. Furthermore, this asymmetry exists primarily because of policies maintained by developed countries who could potentially be flooded with foreign workers were they to ease their restrictions on the movement of workers. Yet on purely economic grounds, the case for the free mobility of labour is no weaker than that for capital. One major reason why multilateral negotiations on investment are unlikely to be successful is that developed countries would be quite unwilling to accept a large increase in labour mobility in return for greater openness to capital inflows on the part of developing countries. However, barriers to foreign investment have come down primarily because developing countries have come to realise that they too benefit. Investment liberalisation has occurred unilaterally as well as bilaterally. The voluntary nature of this liberalisation makes it difficult to view it as a concession on the part of developing countries.

5.2 ITT and the Mandate of the WTO

Document WT/WGTT/3/Rev.1 states that:

> *It will be useful for the Working Group to consider, inter alia, in its future meetings, steps that might be taken within the mandate of the WTO to increase flows of technology as envisaged in the above provisions, to developing countries in accordance with the mandate contained in Paragraph 37 of the Doha Ministerial Declaration.*

As has been argued above, adequate openness to the rest of the world is needed to allow the process of international ITT to take place. Even countries such as Japan and Korea, who were not so welcoming toward inward FDI during certain phases of their development, were rather keen on international trade. Thus, the traditional focus of the WTO – the creation and maintenance of an open world trade regime – complements the goals of the WGTT.

As noted earlier, the evidence shows how FDI results in ITT, even though it does not always lead to spillovers for local competitors of multinational firms. While there are major multilateral agreements (GATT and the GATS) that deal with international trade in goods and services, FDI flows are not subject to multilateral rules. Given the intimate connection between FDI and ITT, any multilateral disciplines on FDI would have direct implications for ITT. However, at the present time, the overall case for a multilateral investment agreement is not obvious. While there are several arguments in favour of such an agreement, practical realities, as well as the limited ground covered in services liberalisation via the GATT, suggest that the time for an investment agreement is not ripe.

Writing and implementing explicit multilateral rules on ITT via FDI is a daunting task. First, how does one design rules for ITT when much of what is to be transferred is private information? By its very nature, ITT involves parties that are asymmetrically informed about the good (i.e. technology) that is being traded. If the buyers themselves lack information, is it reasonable to assume that a third party (such as the local government of a developing country) would have the necessary information to implement a socially beneficial policy? To put it even more directly, rules ought to be based on things that can be easily quantified – but technology is difficult to measure. Royalty payments can give some ideas about market-mediated technology transfer but, due to the problem of transfer pricing, they cannot be trusted to give an accurate indication of the ITT that takes place between multinational firms and their subsidiaries.

The information problem described above is just one hurdle facing multilateral disciplines on ITT. Yet another fundamental problem is that ITT requirements will not be credible when a developing country is desperate to attract investment. An ITT requirement may work when a large country is seeking investment into a project from several interested investors. However, most developing countries (and especially the least developed ones) will not find themselves in such a situation. Since such countries can hardly refuse investment projects that do not bring adequate ITT (as perceived by the host country), any ITT requirement would be a moot issue since firms could easily refuse to comply with conditions they find too cumbersome, well knowing that the project would be ultimately approved.

Another practical reality confronting potential multilateral disciplines on technology transfer is that such disciplines would work very much like TRIMS; the existence of the TRIMS agreement proves that such disciplines are not acceptable to the

WTO. The same parties that oppose the use of TRIMS are likely to oppose the use of ITT requirements (and for similar reasons). Thus, the practical realities of the WTO make it unlikely that any ITT requirements would actually come into place.

Thus the case for multilateral disciplines on technology transfer seems rather weak. Does this mean that multilateral co-operation on ITT at the WTO is an endeavour with dim prospects? This is also not the case. While any multilateral disciplines on the magnitude of ITT would be difficult do design and enforce, successful multilateral co-operation on ITT can and should occur in their absence. In particular, multilateral co-operation with respect to ITT will be fruitful if it can help correct the distortions that are present in the market for technology. As has been noted, asymmetric information and the lack of perfect competition in the market for technology are fundamental problems. Can these problems be alleviated by multilateral co-operation? If so, how?

5.3 The Role of the WTO in Resolving Information Problems

Multilateral co-operation that can help limit the scope for opportunism on the part of buyers and sellers can improve the process of ITT. Problems of asymmetric information can be resolved by increased collaboration and information sharing between WTO members. Such collaboration needs to occur both among developed countries and with developing countries and it needs to be multi-faceted in nature.

Thus far, the type of information sharing that has occurred at the WGTT via the discussion of country experiences has been of a rather general nature. While helpful, this information exchange does not go far enough. Countries that have been relatively successful in developing technological capability can certainly do more to share their knowledge with those that are struggling to succeed. As an example, consider the case of Japan. It is well documented that the Japanese Ministry of Industry and Trade played an active role in encouraging ITT to Japan. What were the steps that MITI took to encourage ITT to local Japanese firms? While a reasonably good general description of the Japanese experience is available in the existing academic literature, practical details about the policies adopted are not easily available. Official information sharing by successful countries can help developing countries design appropriate policies for ITT.

Improved knowledge about the experience of successful countries would also help to clarify whether today's global environment is conducive to a duplication of policies that have proved successful in the past. For example, those countries that are seeking to access foreign technology via licensing could benefit from knowing explicit details about past licensing contracts. Many developing countries have little knowledge about the structure of international ITT contracts. What are reasonable royalty rates? What sort of conditions have sellers of technology been willing to accept? What types of contract clauses have proved helpful in encouraging local technological development? Answers to such important questions are available but their dissemination requires concerted efforts on the part of the private as well as the public sector of developed countries. Privacy concerns might be raised, but these cannot be all important: past licensing contracts that have already expired can hardly raise serious privacy concerns.

Information sharing of the type described above might be viewed as a type of technical assistance, but it goes further than the typical notion of technical assistance. What is being argued here is that developed countries that underwent rapid technological development can teach developing countries important lessons if they are willing and able to tap the knowledge available in both public and private sectors. Sharing of country experiences that does not utilise the information available to the private sector can only be of limited use. Furthermore, developing countries can play a similar role for the least developed ones: the experiences of countries such as Japan are highly relevant for countries such as Brazil and India who, in turn, can play a similar role for the least developed countries. Finally, once sufficient information is available, it will be possible to address the following practical question: can one design a model technology transfer contract that can help protect the interests of both buyers and sellers? The WTO can do much to encourage the development of such a model which can subsequently be made available to all countries.

To summarise, the argument here is not that the WTO needs to be an intermediary in actual technology transfer transactions, but rather that it should encourage information gathering and sharing which can ensure that such transactions occur smoothly. Once sufficient information is available, one might be able to design a model contract that can alleviate the problems that bedevil the exchange of technology.

5.4 Curtailing Market Power

Holders of patents and other IPRs possess substantial market power. Such market power is the intended result of IPRs – the profits that result from market power are often a just reward for past R&D efforts. However, as has been made obvious by the controversy surrounding the 'appropriate' prices of drugs and medicines in developing countries, market power can raise issues of both equity and efficiency. In fact, the issue has been raised indirectly at the WGTT by the delegation that wondered about the relationship between competition policy and ITT. Since there are no multilateral disciplines in the area of competition policy, this is a fertile area for discussion. The WTO can potentially help prevent abuses of market power with respect to ITT by imposing appropriate multilateral disciplines.

While direct requirements on ITT imposed by host countries (as in TRIMS) are probably undesirable, this is not true of multilateral disciplines that can help correct the distortions in the technology market. However, the difficulty with this prescription is that the adoption of multilateral disciplines on competition is a much broader issue and considerations other than market power abuse by holders of intellectual property are involved. In fact, most developing countries (and especially the least developed ones) are not in a position to implement and enforce competition laws against large multinational firms. Here again, the developed countries can partly assist the process of technology transfer by enforcing competition law on behalf of developing countries. In fact, a committed effort on the part of developed countries to prevent market power abuses by their sellers of technology can do much to achieve the goals of Article 66.

5.5 The TRIMS Agreement

In the Uruguay Round, an agreement on TRIMS was negotiated. This agreement prohibits measures that are inconsistent with national treatment and the GATT ban on the use of quantitative restrictions. It contains an illustrative list of prohibited measures, including local content, trade balancing, foreign exchange balancing and domestic sales requirements. Furthermore, it requires that all non-conforming policies be notified within 90 days of entry into force of the agreement and that these be eliminated within two, five or seven years, for industrialised, developing and least developed countries, respectively. The TRIMS agreement did not go much beyond existing GATT rules – it simply reiterated the GATT national treatment principle and the prohibition of quantitative restrictions in the context of certain investment policies that are deemed to be trade related. The GATT has been a constraint on countries using TRIMS, and can be expected to become a more serious source of discipline in the future as Uruguay Round transition periods for developing countries expire.

If domestic distortions and externalities from FDI are absent, then governments should allow for unfettered market transactions with respect to FDI. For example, under perfect competition, domestic content protection lowers welfare by raising the price of domestic inputs: the resulting benefits to input suppliers are outweighed by the costs incurred by final goods producers. A rationale for policies restricting FDI can arise if domestic policy distortions or market failures exist. Since multinational firms typically arise in oligopolistic industries, the presence of imperfect competition in the host economy is an obvious candidate. Analyses of content protection and export performance requirements under conditions of imperfect competition illustrate that the welfare effects of such policies need not be always negative. However, the standard normative prescription applies: more efficient instruments can be identified to address the specific distortion at hand. For example, in the case of anti-competitive practices resulting from market power or collusion, appropriate competition policies need to be used. Similarly, domestic policy distortions such as tariffs should be removed at source. If the distortion is due to some type of market failure, an appropriately designed regulatory intervention is required. Further, such intervention needs to be applied on a non-discriminatory basis to both foreign and domestic firms. This approach is implicit in the WTO, which not only aims at progressive liberalisation of trade, but also imposes national treatment and most favoured nation constraints on policies. The adoption of such principles entails a prohibition on the use of most trade-related investment measures (TRIMS).

The literature on TRIMS also notes that there may indeed be circumstances where, from the viewpoint of an individual country, its optimal second-best policy toward inward FDI has a restrictive flavour. However, such policies typically have a beggar-thy-neighbour effect. If all countries pursue such policies, the outcome will be inefficient from a world welfare point of view. Under such circumstances, co-operation that involves agreement not to restrict FDI can be Pareto improving. Alternatively, the situation may be zero-sum, in which case there are no gains from co-operation.

Surveys often find that investment measures require firms to take actions that they would have taken anyway. For example, a policy that requires firms to export is inconsequential if firms find it advantageous to export even in the absence of such a requirement. In other words, existing survey research shows that TRIMS often fail to bind. However, these surveys did not take account of the firms that may have refrained from investing in countries with TRIMS. By discouraging FDI and distorting the allocation of global capital, the use of TRIMS by an individual country may have efficiency consequences for the world.

Whatever the economic rationale behind TRIMS, the available empirical evidence suggests that local content and related policies are costly to the economy. Furthermore, they often do not achieve the desired backward and forward linkages, encourage inefficient foreign entry and create potential problems for future liberalisation: those who successfully enter a market when it is subject to some investment measures lobby against a change in regime. However, existing evidence on the efficacy of TRIMS does not really examine whether domestic content requirements might play a role in the technological development of local firms. In short, while there may be a connection between TRIMS and technological development, research has not shed light on the issue in either direction.

References

Aghion, Philippe and Peter Howitt (1990). 'A Model of Growth through Creative Destruction', *Econometrica* 60: 323–51.

Aitken, Brian, Gordon H. Hanson and Ann E. Harrison (1997). 'Spillovers, Foreign Investment, and Export Behavior', *Journal of International Economics* 43: 103–32.

Aitken, Brian and Ann E. Harrison (1999). 'Do Domestic Firms Benefit from Direct Foreign Investment?', *American Economic Review* 89: 605–18.

Aitken, Brian, Ann E. Harrison and Robert E. Lipsey (1996). 'Wages and Foreign Ownership: A Comparative Study of Mexico, Venezuela and the United States', *Journal of International Economics* 40: 345–71.

Balasubramanyam, Venkataraman N., Mohammed A. Salisu and David Sapsford (1996). 'Foreign Direct Investment and Growth in EP and IS Countries', *Economic Journal* 106: 92–105.

Barrell, Ray and Nigel Pain (1997). 'Foreign Direct Investment, Technological Change, and Economic Growth within Europe', *Economic Journal* 107: 1770–86.

Barry, Frank and John Bradley (1997). 'FDI and Trade: The Irish Host Country Experience', *Economic Journal* 107: 1798–1811.

Batra, Geeta and Hong W. Tan (2002). 'Inter-Firm Linkages and Productivity Growth in Malaysian Manufacturing', Mimeo, International Finance Corporation, World Bank, Washington, DC.

Bayoumi, Tamim, David T. Coe and Elhanan Helpman (1999). 'R&D Spillovers and Global Growth', *Journal of International Economics* 47: 399–428.

Blalock, Garrick (2001). 'Technology from Foreign Direct Investment: Strategic Transfer Through Supply Chains,' Mimeo, University of California, Berkeley.

Blomström, Magnus (1986). 'Foreign Investment and Productive Efficiency: The Case of Mexico', *Journal of Industrial Economics* 15: 97–110.

Blomström, Magnus and Hakan Persson (1983). 'Foreign Investment and Spillover Efficiency in an Underdeveloped Economy: Evidence from the Mexican Manufacturing Industry', *World Development* 11: 493–501.

Blomström, Magnus and Ari Kokko (1998). 'Multinational Corporations and Spillovers', *Journal of Economic Surveys* 12: 247–77.

Blomström, Magnus and Fredrik Sjoholm (1999). 'Technology Transfer and Spillovers: Does Local Participation with Multinationals Matter?', *European Economic Review* 43: 915–23.

Blonigen, Bruce A. (1999). 'In Search of Substitution between Foreign Production and Exports', *Journal of International Economics* 53: 81–104.

Bond, Eric and Larry Samuelson (1986). 'Tax Holidays as Signals', *American Economic Review* 76: 820–826.

Borensztein, E., J. De Gregorio and J-W. Lee (1998). 'How does foreign direct investment affect economic growth?', *Journal of International Economics* 45: 115–135.

Branstetter, Lee (2001). 'Are Knowledge Spillovers International or Intranational in Scope? Microeconometric Evidence from US and Japan', *Journal of International Economics* 53: 53–79.

Caves, Richard E. (1974). 'Multinational Firms, Competition, and Productivity in Host-Country Industries', *Economica* 41: 176–93.

Caves, Richard E. (1996). *Multinational Enterprise and Economic Analysis*. Cambridge: Cambridge University Press.

Child, John, David Faulkner and Robert Pithkethly (2000). 'Foreign Direct Investment in the UK 1985–1994: The Impact on Domestic Management Practice', *Journal of Management Studies* 37: 142–166.

Clerides, S. K., Saul Lach and James R. Tybout (1998). 'Is Learning by Exporting Important? Micro-Dynamic Evidence from Colombia, Mexico, and Morocco', *Quarterly Journal of Economics* 113: 903–48.

Coe, David T. and Elhanan Helpman (1995). 'International R&D Spillovers', *European Economic Review* 39: 859–87.

Coe, David T., Elhanan Helpman and Alexander W. Hoffmaister (1997). 'North–South R&D Spillovers', *The Economic Journal* 107: 13–149.

Coe, David T. and Alexander W. Hoffmaister (1999). 'Are There International R&D Spillovers among Randomly Matched Trade Partners? A Response to Keller', IMF Working Paper No. WP/99/18, International Monetary Fund, Washington, DC.

De Long, Bradford J. and Lawrence H. Summers (1991). 'Equipment Investment and Economic Growth', *Quarterly Journal of Economics* 106: 445–502.

Devereux, Michael and Rachel Griffiths (1998). 'Taxes and the Location of Production: Evidence from a Panel of US States', *Journal of Public Economics* 68: 335–67.

Dinopoulos, Elias and Paul Segerstrom (1999). 'The Dynamic Effects of Contingent Tariffs', *Journal of International Economics* 47: 191–222.

Dixit, Avinash K. and Joseph E. Stiglitz (1977). 'Monopolistic Competition and Optimum Product Diversity', *American Economic Review* 67: 297–308.

Djankov, Simeon and Bernard Hoekman (2000). 'Foreign Investment and Productivity Growth in Czech Enterprises', *World Bank Economic Review* 14(1): 49–64.

Dollar, David (1992). 'Outward Oriented Developing Economies Really Do Grow More Rapidly: Evidence from 95 LDCs, 1976–85', *Economic Development and Cultural Change* 523–44.

Eaton, Jonathan and Samuel Kortum (1996). 'Trade in Ideas: Patenting and Productivity in the OECD', *Journal of International Economics* 40: 251–78.

Ethier, Wilfred J. (1982). 'National and International Returns to Scale in the Modern Theory of International Trade', *American Economic Review* 72:389–405.

Ethier, Wilfred. J. and James R. Markusen (1991). 'Multinational Firms, Technology Diffusion and Trade', *Journal of International Economics* 41: 1–28.

Ferrantino, Michael J. (1993). 'The Effect of Intellectual Property Rights on International Trade and Investment', *Weltwirtschaftliches Archive* 129: 300–31.

Findlay, Ronald (1978). 'Relative Backwardness, Direct Foreign Investment, and the Transfer of Technology: A Simple Dynamic Model', *Quarterly Journal of Economics* 62: 1–16.

Finger, Michael J. (1993). *Antidumping*. Ann Arbor, University of Michigan Press.

Gershenberg, Irving (1987). 'The Training and Spread of Managerial know-how: A Comparative Analysis of Multinational and Other Firms in Kenya', *World Development* 15: 931–39.

Glass, Amy J. and Kamal Saggi (1998). 'International Technology Transfer and the Technology Gap', *Journal of Development Economics* 55: 369–98.

—— (2002a). 'Intellectual Property Rights and Foreign Direct Investment', *Journal of International Economics* 56: 387–410.

—— (2002b). 'Licensing versus Direct Investment: Implications for Economic Growth', *Journal of International Economics* 56: 131–153.

—— (2002c). 'Multinational Firms and Technology Transfer', *Scandinavian Journal of Economics* 104: 495–513. Also issued as World Bank Policy Research Working Paper No. 2067, World Bank, Washington, DC.

Globerman, Steve (1979). 'Foreign Direct Investment and 'Spillover' Efficiency Benefits in Canadian Manufacturing Industries', *Canadian Journal of Economics* 12(Feb.): 42–56.

Gould, David M. and William C. Gruben (1996). 'The Role of Intellectual Property Rights in Economic Growth', *Journal of Development Economics* 48: 323–350.

Grossman, Gene M. (1981). 'The Theory of Domestic Content Protection and Content Preference', *The Quarterly Journal of Economics* XCVI: 583–603.

Grossman, Gene M. and Elhanan Helpman (1991). *Innovation and Growth in the Global Economy*, Cambridge, MA: MIT Press.

—— (1995). 'Technology and Trade' in Gene Grossman and Kenneth Rogoff (eds), *Handbook of International Economics*, Vol. 3, Elsevier Science.

Guisinger, Steven et al. (1985). *Investment Incentives and Performance Requirements*. New York: Praeger.

Haddad, Mona and Ann Harrison (1993). 'Are there Positive Spillovers from Direct Foreign Investment? Evidence from Panel Data for Morocco', *Journal of Development Economics* 42: 51–74.

Haskel, Jonathan E., Sonia Pereira and Matthew J. Slaughter (2002). 'Does Inward Foreign Direct Investment Boost the Productivity of Domestic Firms?', NBER Working Paper No. 8724.

Helpman, Elhanan (1993). 'Innovation, Imitation, and Intellectual Property Rights', *Econometrica* 61: 1247–80.

Hines, James R. (1996). 'Altered States: Taxes and the Location of FDI in America', *American Economic Review* 86: 1076–1094.

Hobday, Michael (1995). *Innovation in East Asia: The Challenge to Japan*. Cheltenham: Edward Elgar.

Hoekman, Bernard M. and Michael M. Kostecki (2001). *The Political Economy of the World Trading System*. New York: Oxford University Press.

Hoekman, Bernard and Kamal Saggi (2000). 'Multilateral Disciplines for Investment-Related Polices?' in Paolo Guerrieri and Hans-Eckart Scharrer (eds), *Global Governance, Regionalism, and the International Economy*. Baden-Baden: Nomos-Verlagsgesellschaft.

Hollander, Abraham (1987). 'Content Protection and Transnational Monopoly', *Journal of International Economics* 23: 283–297.

Holmes, Thomas (1998). 'The Effect of State Policies on the Location of Manufacturing: Evidence from State Borders', *Journal of Political Economy* 106: 667–705.

Hsieh, Chang-Tai (2002). 'What Explains the Industrial Revolution in East Asia? Evidence from the Factor Markets', *American Economic Review*, forthcoming.

Huizinga, Harry (1995). 'Taxation and the Transfer of Technology by Multinational Firms', *Canadian Journal of Economics* 28: 648–55.

Irwin, Douglas A. and P. J. Klenow (1994). 'Learning by Doing Spillovers in the Semiconductor Industry', *Journal of Political Economy* 102: 1200–27.

Jones, Charles (1995a). 'Time Series Tests of Endogenous Growth Models', *Quarterly Journal of Economics* 110: 495–525.

—— (1995b). 'R&D-Based Models of Economic Growth', *Journal of Political Economy* 103: 759–84.

Kabiraj, Tarun and Sugata Marjit (1993). 'International Technology Transfer under Potential Threat of Entry', *Journal of Development Economics* 42: 75–88.

Keesing, Donald B. (1982). 'Exporting manufactured consumer goods from developing to developed economies: marketing by local firms and effects of developing country policies'. World Bank, Washington, DC.

Keller, Wolfgang (1996). 'Absorptive Capacity: On the Creation and Acquisition of Technology in Development', *Journal of Development Economics* 49: 199–227.

—— (1998). 'Are International R&D Spillovers Trade-related? Analyzing Spillovers among Randomly Matched Trade Partners', *European Economic Review* 42: 1469–81.

Kinoshita, Yuko and Ashoka Mody (1997). 'Private and Public Information for Foreign Investment Decision', World Bank Policy Research Working Paper No. 1733, World Bank, Washington, DC.

Krugman, Paul R. (1979). 'A Model of Innovation, Technology Transfer, and the World Distribution of Income', *Journal of Political Economy* 87: 253–66.

Lai, Edwin L.C. (1998). 'International Intellectual Property Rights Protection and the Rate of Product Innovation', *Journal of Development Economics* 55: 131–151.

Lai, Edwin L.C. and Larry D. Qiu (1999). 'Northern Intellectual Property Rights Standard for the South?', *Journal of International Economics*, forthcoming.

Lall, Sanjaya (1980). 'Vertical Inter-Firm Linkages in LDCs: An Empirical Study', *Oxford Bulletin of Economics and Statistics* 42: 203–6.

Layton, Duane W. (1982). 'Japan and the Introduction of Foreign Technology: A Blueprint for Lesser Developed Countries?', *Stanford Journal of International Law* 18: 171–212.

Lee, Jeong Y. and Edwin Mansfield (1996). 'Intellectual Property Protection and U.S. Foreign Direct Investment', *Review of Economics and Statistics* 78: 181–86.

Lin, Ping and Kamal Saggi (1999). 'Incentives for FDI under Imitation', *Canadian Journal of Economics*, 32: 1275–98.

Lichtenberg, Frank and B. van Pottelsberghe de la Potterie (1998). 'International R&D Spillovers: A Comment', *European Economic Review* 42: 183–1491.

Lichtenberg, Frank and B. van Pottelsberghe de la Potterie (2001). 'Does Foreign Direct Investment Transfer Technology Across Borders?', *Review of Economics and Statistics* 83: 490–497.

Lipsey, Robert E. and Merle Yahr Weiss (1981). 'Foreign Production and Exports in Manufacturing Industries', *Review of Economics and Statistics* 63: 488–94.

—— (1984). 'Foreign Production and Exports of Individual Firms', *Review of Economics and Statistics* 66: 304–7.

Lucas, Robert E., Jr. (1988). 'On the Mechanics of Economic Development', *Journal of Development Economics* 22: 3–42.

Mansfield, Edwin and Anthony Romeo (1980). 'Technology Transfer to Overseas Subsidiaries by U.S. Based Firms', *Quarterly Journal of Economics* 95: 737–49.

Mansfield, Edwin, Mark Schwartz and Samuel Wagner (1981). 'Imitation Costs and Patents: An Empirical Study', *Economic Journal* 91: 907–18.

Mansfield, Edwin (1994). 'Intellectual Property Protection, Foreign Direct Investment, and Technology Transfer', World Bank and International Finance Corporation, Discussion Paper No. 19.

Markusen, James R. (1995). 'The Boundaries of Multinational Enterprises and the Theory of International Trade', *Journal of Economic Perspectives* 9: 169–89.

—— (2000). 'Contracts, Intellectual Property Rights, and Multinational Investment in Developing Countries', *Journal of International Economics* 53: 189–204.

Markusen, James R. and Anthony Venables (1999). 'Foreign Direct Investment as a Catalyst for Industrial Development', *European Economic Review* 43: 335–56.

Maskus, Keith E. (2000). *Intellectual Property Rights in the Global Economy*, Washington, DC: Institute of International Economics.

Maksus, Keith E. and Christine McDaniel (1999). Impacts of the Japanese Productivity System on Productivity Growth. *Japan and the World Economy* 11: 557–574.

Maskus, Keith E. and Mohan Penubarti (1995). 'How Trade-Related are Intellectual Property Rights?', *Journal of International Economics* 39: 227–48.

Matsuyama, Kiminori (1990). 'Perfect Equilibrium in a Trade Liberalization Game', *American Economic Review* 80: 480–92.

Mattoo, Aaditya, Marcelo Olarreaga and Kamal Saggi (2003). 'Mode of Foreign Entry, Technology Transfer and FDI Policy', *Journal of Development Economics*, forthcoming.

Mazumdar, Joy (2001). 'Imported Machinery and Growth in LDCS', *Journal of Development Economics* 65: 209–224.

Miyagiwa, Kaz and Yuka Ohno (1995). 'Closing the Technology Gap Under Protection', *American Economic Review* 85: 755–70.

—— (1999). 'Credibility of Protection and Incentives to Innovate', *International Economic Review* 40: 143–64.

Moran, Theodore (1998). *Foreign Direct Investment and Development*, Washington DC: Institute for International Economics.

Nadiri, M.I. (1991). 'U.S. Direct Investment and the Production Structure of the Manufacturing Sector in France, Germany, Japan, and the UK', NBER Working Paper.

Nelson, Richard R., and Howard Pack (1999). 'The Asian Miracle and Modern Growth Theory', *Economic Journal* 109: 416–36.

Ozawa, Terutomo (1974). *Japan's Technological Challenge to the West, 1950–1974: Motivation and Accomplishment*, Cambridge, Massachusetts: The MIT Press.

Pack, Howard (1992). 'Technology Gaps Between Industrial and Developing Countries: Are there Dividends for Late-comers?', Proceedings of the World Bank Annual Conference on Development Economics, Supplement to the World Bank Economic Review and World Bank Research Observer, World Bank: Washington, DC (pp. 283–302).

—— (1994). 'Endogenous Growth Theory: Intellectual Appeal and Empirical Shortcomings', *Journal of Economic Perspectives* 8: 55–72.

—— (1997). 'The Role of Exports in Asian Development' in Nancy Birdsall and Frederick Jaspersen (eds), *Pathways to Growth: Comparing East Asia and Latin America*, Washington, DC: Inter-American Development Bank.

Pack, Howard and Kamal Saggi (1997). 'Inflows of Foreign Technology and Indigenous Technological Development', *Review of Development Economics* 1: 81–98.

—— (2001). 'Vertical Technology Transfer via International Outsourcing', *Journal of Development Economics* 65: 389–415.

Pack, Howard and Larry E. Westphal (1986). 'Industrial Strategy and Technological Change: Theory versus Reality', *Journal of Development Economics* 22: 87–128.

Parente, Stephen L. and Edward C. Prescott (1994). 'Barriers to Technology Adoption and Development', *Journal of Political Economy* 102: 298–321.

Pritchett, Lant (1997). 'Divergence, Big Time', *Journal of Economic Perspectives* 11: 3–17.

Ramachandran, Vijaya (1993). 'Technology Transfer, Firm Ownership, and Investment in Human Capital', *Review of Economics and Statistics* 75: 664–70.

Rhee, Yung Whee (1990). 'The Catalyst Model of Development: Lessons from Bangladesh's Success with Garment Exports', *World Development* 18: 333–46.

Rhee, Yung Whee, Bruce Ross-Larson and Gary Pursell (1984). *Korea's Competitive Edge: Managing Entry into World Markets*. Johns Hopkins, Baltimore.

Rivera-Batiz, Louis A. and Paul Romer (1991). 'Economic Integration and Endogenous Growth', *Quarterly Journal of Economics* 106: 531–56.

Roberts, Mark and James Tybout (1997). 'The Decision to Export in Columbia: An Empirical Model of Entry with Sunk Costs', *American Economic Review* 87: 545–64.

Rodriguez, Francisco and Dani Rodrik (1999). 'Trade Policy and Economic Growth: A Skeptic's Guide to the Cross-country Evidence', NBER Working Paper No. 7081, National Bureau of Economic Research, Boston, Mass.

Rodriguez Clare, Andres (1996). 'Multinationals, Linkages, and Economic Development', *American Economic Review* 86: 852–74.

Rodrik, Dani (1987). 'The Economics of Export-Performance Requirements', *The Quarterly Journal of Economics* 102: 633–650.

Romer, Paul (1990). 'Endogenous Technological Change', *Journal of Political Economy* 98: S71–S102.

—— (1993). 'Idea Gaps and Object Gaps in Economic Development', *Journal of Monetary Economics* 32: 543–73.

Sachs, Jeffrey and Andrew Werner (1995). 'Economic Reform and the Process of Global Integration', *Brookings Papers on Economic Activity* 1: 1–118.

Saggi, Kamal (1996). 'Entry into a Foreign Market: Foreign Direct Investment versus Licensing', *Review of International Economics* 4: 99–104.

—— (1999). 'Foreign Direct Investment, Licensing, and Incentives for Innovation', *Review of International Economics* 7: 699–714.

—— (2002). 'Backward Linkages under Foreign Direct Investment', mimeo, Southern Methodist University.

Sakong, Il (1993). *Korea in the World Economy*, Washington DC: Institute for International Economics.

Schiff, Maurice, Yanling Wang and Marcelo Olarreaga (2002). 'Trade-Related Technology Diffusion and the Dynamics of North–South and South–South Integration', Mimeo, World Bank, Washington, DC.

Segerstrom, Paul S., T. C. A. Anant and Elias Dinopoulos (1990). 'A Schumpeterian Model of the Product Life Cycle', *American Economic Review* 80: 1077–91.

Smarzynska, Beata K. (1999). 'Composition of Foreign Direct Investment and Protection of Intellectual Property Rights in Transition Economies', Mimeo, World Bank, Washington, DC.

—— (2002). 'Spillovers from Foreign Direct Investment through Backward and Forward Linkages: The Case of Lithuania', Mimeo, World Bank, Washington, DC.

Smith, Pamela (1999). 'Are Weak Patent Rights a Barrier to U.S. Exports', *Journal of International Economics* 48: 151–77.

Taylor, M. Scott (1994). 'TRIPS, Trade and Growth', *International Economic Review* 35: 361–81.

Teece, David J. (1977). 'Technology Transfer by Multinational Firms: The Resource Cost of Transferring Technological Know-how', *Economic Journal* 87: 242–61.

Tybout, James and M. Daniel Westbrook (1995). 'Trade Liberalization and Dimensions of Efficiency Change in Mexican Manufacturing Industries', *Journal of International Economics* 39: 53–78.

UNCTAD (1992). *World Investment Report: Transnational Corporations as Engines of Growth*, New York: United Nations.

—— (1997). *World Investment Report: Transnational Corporations, Market Structure, and Competition Policy*, New York, NY: United Nations.

—— (1998). *World Investment Report: Trends and Determinants*, New York, NY: United Nations.

—— (1999). *World Investment Report: Foreign Direct Investment and the Challenge of Development*, New York, NY: United Nations.

—— (2000). *World Investment Report: Cross Border Mergers and Acquisitions and Development*, New York: United Nations.

—— (2001). *World Investment Report: Promoting Linkages*, New York: United Nations.

United Nations (1983–98). *International Trade Statistics Yearbook*, New York: United Nations.

UNCTC (1991). *Government Policies and Foreign Direct Investment*. United Nations, New York, NY.

Vishwasrao, Sharmila (1995). 'Intellectual Property Rights and the Mode of Technology Transfer', *Journal of Development Economics* 44: 381–402.

Wang, Jian-Ye and Magnus Blomström (1992). 'Foreign Investment and Technology Transfer', *European Economic Review* 36: 137–55.

World Bank (1999). *World Development Indicators*, Washington, DC: World Bank.

World Trade Organization (1998). 'Report of the Working Group on the Relationship Between Trade and Investment to the General Council', WT/WG/TI/2, 8 December.

Xu, Bin (2000). 'Multinational Enterprises, Technology Diffusion, and Host Country Productivity Growth', *Journal of Development Economics* 62: 477–93.

Xu, Bin and Jianmao Wang (1999). 'Capital Goods Trade and R&D Spillovers in the OECD', *Canadian Journal of Economics* 32: 1258–74.

—— (2000). 'Trade, FDI, and International Technology Diffusion', *Journal of Economic Integration* 15: 585–601.

Yang, Guifang and Keith E. Maskus (2001). 'Intellectual Property Rights, Licensing, and Innovation in an Endogenous Product-Cycle Model', *Journal of International Economics* 53: 169–88.

Young, Alwyn (1995). 'The Tyranny of Numbers: Confronting the Statistical Realities of the East Asian Growth Experience', *Quarterly Journal of Economics* 110: 641–80.